THE EYE

THE EYE

A Natural History

SIMON INGS

BLOOMSBURY

First published in Great Britain 2007

Bloomsbury Publishing Plc,
36 Soho Square,
London W1D 3QY

A CIP catalogue record for this book
is available from the British Library

ISBN 9780747578055

10 9 8 7 6 5 4 3 2

Typeset by Hewer Text UK Ltd, Edinburgh
Printed by Clays Ltd, St Ives plc

Bloomsbury Publishing, London, New York, Berlin

The paper this book is printed on is certified by the © 1996 Forest
Stewardship Council A.C. (FSC). It is ancient-forest friendly. The printer
holds FSC chain of custody SGS-COC-2061

FSC
Mixed Sources
Product group from well-managed
forests and other controlled sources
Cert no. SGS-COC-2061
www.fsc.org
© 1996 Forest Stewardship Council

www.bloomsbury.com/simonings

For Natalie

– new eyes for old

CONTENTS

ACKNOWLEDGEMENTS

For years, my agent Peter Tallack has been looking for someone to write a popular, single-volume work about the eye. For years, we gabbled about the project. His immense enthusiasm and patience have made this book possible. For their guidance and encouragement I would also like to thank Philip Ball, Michael Land, Oliver Morton and Dan-Eric Nilsson. At Bloomsbury, Rosemary Davidson, Rosemarie Hudson, Mary Instone and Mike Jones have worked very hard over the years to keep both book and author coherent. Closer to home I would like to thank my wife, Anna, for her support, encouragement, editing, and bill-paying. Daniel Brown, Marcus Chown, Stuart Clark and Mateo Willis provided me with countless valuable leads and ideas.

For their permission to quote passages from the following works, I am indebted to Richard Gregory (*Recovery from Early Blindness* by R. Gregory and J. Wallace, 1963), W. W. Norton & Co. Inc. (*Visual Intelligence* by D. Hoffman, 1998) and Simon & Schuster (*Eye and Camera* by G. Wald, 1953).

I would also like to thank all those who granted me permission to reproduce their figures and photographs. Most are credited on p. 311. In addition, for the following illustrations, I would like to thank Elsevier (57); Harvard News Office (68); John Wiley & Sons, Inc. (36); New York Academy of Medicine (59); the NIH Record and the National Institutes of Health (77); Springer, Berlin (108). I would also like to thank the BOA Libarary at the College of Optometrists for sourcing the illustrations on pp. 1, 5, 14, 52, 71, 73, 102, 103, 128, 170, 180, 181, 190, 193, 200, 211, 232, 244, 254, 264, 286, 291 and Plates 10 and 16; the Cranbrook Institute of Science for the drawings by G.L. Walls on pp. 5, 73, 254 and 255; Henry Holt & Co. for permission to reproduce the photograph of tissue stained by the Golgi method on Plate 9; the *International Journal of Development Biology* for the images of Walter Gehring's fruit fly experiments on Plate 12, and for the photograph on p. 82; Dan-Eric Nilsson for permission to adapt the figure on p. 98; the Oxford University Press for permission to reproduce a figure (p. 20) and a photograph (the pineal organ on Plate 11) from Jerome Wolken's *Light Detectors, Photoreceptors, and Imaging Systems in Nature* (1995); the Royal Society of Edinburgh for David Brewster's diagram on Plate 3; the San Francisco Museum of Modern Art for Sigmund Exner's photograph through a firefly eye on Plate 2; and Nicholas Wade for the portrait of Jan Purkinje on p. 40.

PROLOGUE

Youth and age

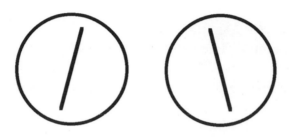

Natalie, my daughter, began life with one eye orbit placed centrally in her forehead. It appeared early – barely a week after her conception – in October 2002 and had things gone awry, there it would have remained, a wet cyclopean hollow, glowering in the shadow of a grotesque proboscis that would have grown in place of her nose.[1]

There are many explanations for what inspired the legend of the Cyclopes – the one-eyed monsters of classical mythology – invoking everything from an ancient find of mysterious dwarf elephant skulls, to the blacksmith's habit, in the days before protective goggles, of protecting one eye with a patch. But the most likely inspiration is also the saddest; occasionally, a human cyclops survives in the womb long enough to emerge visibly disfigured. Once upon a time, someone got a good look at what they had aborted – and for ever after wished they hadn't.

This happened not so long ago. George Gould and Walter Pyle's indispensable volume *Anomalies and Curiosities of Medicine* of 1896 includes an account of a woman of thirty-five, the mother of seven children, 'who while pregnant was feeding some rabbits, when one of the animals jumped at her with its eyes "glaring" upon her, causing a sudden fright.'[2] The child's mouth and face were small and rabbit-shaped. Instead of a nose, it had 'a fleshy growth 3/4 inch long by 1/4 inch broad, directed upward at an angle of 45 degrees.' Between the nose and the mouth was an organ resembling an adult eye. Within it lay 'two small, imperfect eyes which moved freely while life lasted.'

The disfigurement of the face was the outward mark of an even more collosal failure; the failure of the brain to divide into two delicately linked halves. So, mercifully, after about ten minutes the child died. The mother went on to have two more, perfectly healthy, children.

Human beings arise, not from dust, but from stuff hardly more edifying: a gellid spittle. How, from this unpromising material, anything beautiful emerges – let alone anything that comprehends and communicates something of the world beyond itself – is a mystery too big to be encompassed by just one version of events. How we explain how we grow depends largely on fashion. The current fashion is to talk of rules embedded, like a code, in every cell. But this, like most shorthand explanations, fosters misunderstanding. Natalie was not fashioned by rules. Nothing oversaw her growth. Genes set the conditions of the game, but the game itself built her: the touch and slide of surfaces, the little chemical kisses, the partings and the reunions, each part of her tissue communicating chemically with each neighbouring part as she folded herself into being. This dance, known to biology as induction, told her brain to split, squeezed her single orbit into two, and drew two little strings of brain out to the windows in her skull to make her eyes.

When a newspaper announces that someone has found the gene 'for' something or other, the sentiment is not so much wrong as insufficient. Every gene is 'for' a virtually infinite number of things, as vital to the whole – and as meaningless in itself – as a single key on a piano keyboard. It is the performance that matters. The chances that a woman grows a cyclops in her womb because of a genetic defect is very slim. It is much more likely that the performance was interrupted – that, for instance, she ate cow cabbage (*Veratrum californicum*, a notorious poisoner of hungry sheep and cattle) early in her pregnancy.

Anna, my wife, has never been so hungry that she has been tempted to eat cow cabbage but necessity did lead us to throw every toxin known to the construction trade at little Natalie. The beautiful ramshackle house we moved into, six months before her birth, chose the moment of our arrival to give up the will to live. Faced with bust plumbing and falling ceilings, we had a stark choice: either we shored the place up, then and there, or we abandoned it. There was no middle way. So, we knocked down walls and filled the air with horse-hair and lime. We sawed up MDF panels without wearing masks. (In the United States they insist you wear protection before handling the stuff.) We stripped the lead-based paint from doors, with acid paste. We sanded boards and varnished them with polyurethane. We sprayed for woodworm, cockroaches, and larger vermin. With her incoming nourishment filtered clean by her mother's placenta, Natalie was not in the least troubled, and by November her eyes had acquired all their basic anatomy.

There is light in the womb – enough to need managing. Natalie's irises – plates of coloured muscle that respond readily to light – were still very young, so her eyelids grew and fused together, shielding her eyes for a while. The following May, with the irises all but fully grown, the eyelids parted, and Natalie blinked.

By April, the centre of each iris had withered, leaving a little hole or 'pupil' for her to see through. There were still three months to go before her birth but Natalie could already see.

Many of the parts of the eye were first named by Imperial Rome's

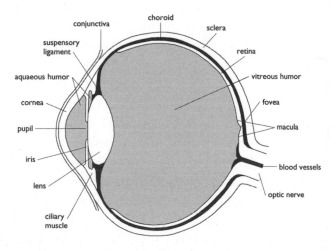

The human eye.

most celebrated anatomist, Claudius Galenus of Pergamum (Galen, to English speakers). Other anatomical names, which sound as if they might have been his, turn out to have been cobbled together quite recently. As a lingua franca for exact study, the cod Graeco-Latin hotchpotch assembled by the biological sciences is a master-piece of precision – it could not have been designed better. Even now, at least for English speakers, the words are not hard to under-stand, thanks to the grab-bag nature of English – this little island's voracious tongue, that swallows invading languages whole.

When Galen described the eye, he hit upon a great analogy: the layers of the eye are like layers of clothing – *tunicae*, or tunics. Working inwards, he described the eyeball just as you might describe what someone is wearing. The analogy is a good one, because no layer of clothing is shaped exactly like another, nor is it necessarily made of the same stuff. My jacket covers my shirt, but not completely – the first couple of shirt buttons are visible, and my shirt cuffs peep below the jacket's sleeves. It is also made of different stuff; it looks different, and hangs differently.

Galen's analogy is casual, and when he talks about one tunic of the eye 'surrounding' another, we need not take him too literally. The eye is not a geometrical figure, sealed off from the rest of the body. Like everything else, it has its exits and its entrances, is folds and frills.

The outermost layer – the white of the eye – is the *sclera*. This is not Galen's word, but an English invention, cooked up in 1888 by taking a Greek adjective meaning 'hard', and mangling it to behave like a Latin noun. No one is confused or puzzled by it for long, because the English word 'sclerotic' – older by two centuries than the invented 'sclera' – means 'hardened and inflexible'. (Actually, there was no pressing need for a new word: Galen's word for the sclera, *consolidativa*, is even more suggestive.)

Almost every sclera in the animal world is pigmented. In humans it is white – bluish-white in Natalie's case, being so thin; yellowish in my mother's, as age thickens it year by year.

The sclera is made of collagen – a rather fatuous statement, consid-ering the astonishing variety of forms that collagen can take (four-tenths of our body proteins are collagen of one sort or another).

Collagen makes the white of the eye, but it also makes the transparent cornea, the artist's 'window on the soul', through which we see. The difference lies in the construction. The cornea, like the white, is solid collagen, but in it, the fibres are laid down with such fine geometrical care that it admits light.

Within the sclera lies a middle tunic, the *uvea*. This word is (almost) one of Galen's. (His term was 'uva', meaning a bunch of grapes. Around 1525, for no reason, someone added an 'e'.) Open an eyeball out at the pupil, and you will find that 'bunch of grapes' is one of the most telling and poetic of all anatomical descriptions. The whole inside structure of the eye is veined with translucent blood vessels, strongly reminiscent of the vessels running through a peeled grape. And the tissue immediately around the pupil, which feeds and operates the brightly pigmented iris, has a clustered appearance.

Galen broke the uvea down into several parts, the better to explain what he saw. One blood-filled, semi-translucent layer, which ran across the back of the eye, looked to him like a 'chorion' – the membrane

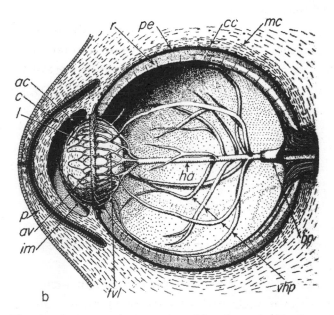

The developing eye's temporary blood supply (the hyaloid),
from Gordon Lynn Walls' mammoth 1942 volume,
The Vertebrate Eye and its Adaptive Radiation.
For more on Walls, see chapter nine.

sac which surrounds the foetus. He dubbed this layer the *choroid*, and the name has stuck.

The choroid is full of blood, feeding the tissues of the eye (much of the eye consists of tiny blood vessels) and we now know what Galen could not – that in the foetus, the choroid gives rise to all the other structures of the uvea. As her months in the uterus passed, the choroids of Natalie's eyes waxed and waned. Some became the pretty circles around each pupil, rings of muscle called irises (meaning 'rainbows', another of Galen's happy neologisms). As it dilates and contracts, the iris focuses the view, guards against too strong a light, and even signals feeling, as it widens the pupil to accommodate a loved one's face. Some of my daughter's choroid became the muscle that helps her focus, stretching her crystalline lenses to adjust their power. Some turned black, coating the back of the eye – rather like the backing-plate of a camera – with a heavily pigmented layer, one cell thick.

By May 2003, the blood vessels of Natalie's choroid layer had withdrawn from her lenses. These no longer required a blood supply, for they were (but for a single layer, one cell thick) quite dead. Isolated from neighbouring tissues only five weeks into gestation, afloat in a watery suspension, the lens is a tissue uniquely isolated from the rest of the body – an exquisitely arranged meshwork of dead cells packed with clear proteins called crystallins which, though soluble, remain virtually unchanged throughout life.

The choroid, as it withdrew from Natalie's lenses, now folded itself back over the eye's 'backing plate' to form the third and final of Galen's tunics – the *retina*. 'Retina' means mesh or net, and when Galen used the word, he was referring to the fine layer of blood vessels lying across the 'backing plate' of the eye. For him, the retina was just a thin, stretched extension of the uvea. Today, we use the word in quite a different sense, to describe the most mysterious of the eye's tunics: the layer that sees. It was a layer of which Galen knew nothing: it was quite invisible to him, and remained undetected until 1834. To call it a 'retina' is, strictly speaking, a misnomer: the modern 'retina' is not at all what Galen was writing about, and is not remotely like a net. It is a cup of transparent nerve tissue, literally an outpost of the brain.

*   *   *

Developing nerve cells shake hands with each other indiscriminately. Their connections acquire a shape and an architecture only when they are stimulated: the busiest are preserved, while the less travelled lines wither.

Natalie's retinal cells were twice taught how best to connect: first by dreams, and then by light.

By April, Natalie had begun, now and again, to sleep. Every so often she closed her eyelids and kept them shut, and when she did, her eyes trembled, stirring the oxygen-rich liquid, the aqueous humour, that lies between the iris and the cornea. (The occasional rapid eye movements that accompany our dreams have at least one very practical purpose: they feed the front of the eye.[3]) The trembling woke the light-sensitive cells of Natalie's retinas. Stimulated, her retinal cells fired at random, preserving and strengthening their connections.[4] Even before she saw, Natalie went through the motions of dreaming, and those motions taught the cells in her retina to hold hands.

Prepared by dreams, Natalie's retinal cells took their next lessons from light. Even before birth, the body is no stranger to illumination. Flesh itself lights up a little, every time a nerve fires.[5] Perhaps this familiarity with light is why any young nerve cell, transported to the retina – the body's most light-sensitive surface – will learn to see, just as every seeing nerve cell, moved elsewhere, adapts to its new environment.[6]

The womb is not dark: it is easily penetrated by light from the outside. From November 2002 to July 2003, the month of her birth, Natalie's retinas seized what news they could through the amniotic murk. They adapted to darkness and to blur. One layer of nerves grew into highly light-sensitive cells, *rods*, superb harvesters of light. At birth, Natalie was well on the way to acquiring good nocturnal vision.

Babies see well at night, but in the glare of day, they are all but blind. A sunny day is a million times brighter than a night-time nursery, and a completely different form of vision is needed to handle such a glare – a form Natalie had not yet acquired. (It is one of the ironies of birth that it fills our world with light and blinds us in the process.)

Exposed to the light, Natalie's eyes had not so much to 'adapt' to the brightness of day, as acquire a new way of seeing. In her retinas lay another set of nerve cells, *cones*. Cones, when they are fully grown, look very different to rods. Indeed, these two kinds of photo-sensitive cell are only distantly related; the distinction between them had arisen in the eyes of jawless fish swimming in Devonian seas around 400 million years ago.[7] At birth, Natalie's cones were little different from ordinary nerve cells, but once exposed to the glare of day they began to change. Natalie will be six before her cones are fully grown, packed tightly into the fovea, the tiny circle on the retina where images are focused and light explodes with colour.

This first year of my daughter's life is visually simple. A few familiar faces. Eyes and smiles. The rest is a blur, slowly resolving as her lenses – dead and round as the lenses of the fish that were her distant ancestors – acquire the slimline shape proper to a land mammal. Though dead, lenses grow: layers of protein are laid on like the skins of an onion, changing their profile and their power. It is a process as natural and unstoppable as the growth of hair and nails. Even now, as my daughter's lenses take on their optimum shape, they are growing thicker, less permeable, and imperceptibly more stiff. This constant accretion of material will eventually spoil her sight, as it is about to spoil mine and as it has, or will, spoil yours. The eye focuses on objects at at different distances by means of a ring of muscle, the ciliary muscle, around the lens. As they harden, the harder the ciliary muscle has to work to wrench our lenses to the right shape. The decline in lens flexibility is smooth and gradual, and we rarely notice it until the process begins to affect the ease with which we can read print.

I look at Natalie through lenses that are forty years old. In a few years, these words will be a blur to me; my lenses will not bend to accommodate them. I will moan, crack some mordant joke about my age, and go to the optician, who will sell me some glasses – lenses hung on wires before the eyes, strangest of all nature's strange variations on the theme of vision. The top half of each lens will allow me to see distant objects, and the lower part will help me see close ones. Benjamin Franklin invented bifocals – glasses whose lenses are, effectively, several lenses in one. This is one of those half-interesting factoids which bring the ageing spectacle-wearer no comfort whatsoever.

Natalie may be short-sighted. If she is, it doesn't matter. A third of all infants are born myopic. The retina – which is to say, the brain – has the problem in hand. It hungers for a clear view, and since it directs the growth of the eyeball, it will make what corrections it can. By the time she is six, we will know whether her aquatic ancestry is so strong that it spoils her sight, leaving her lenses too round and powerful to see the world with clarity. If it is, she will wear glasses to correct her 'short sight'.

For the first couple of months after she was born, as she tried to figure out how to look at the world, Natalie used to cross her eyes a lot. Once she began to pick things up, her hands began to teach her eyes how to see. Her eyes imagined there were two identical blocks before her; her fingers found only one. Her eyes, vanquished, uncrossed.

By ten months old, Natalie sees the world pretty much as it is. She knows what is near, and what is far away. I see her thinking: 'How do I reach that?' She stretches out her arm to me. I am sitting on the other side of the room. She knows I am far away, but she doesn't know how far. She cannot crawl yet, so 'far away' to her simply means 'out of reach'.

Seeing her reach out, I can't help but smile. She returns the smile and reaches out again. I lean forward and reach out for her. We're too far away to touch, but this is our game. We reach out, and smile, and this will be a kind of touch. Mind will brush mind. Impossible to imagine that she thinks any differently; that we are playing different games; that she has no idea, yet, about mind; about what I am, or what she is.

Anna comes into the room. Anna is my wife, Natalie's mother, the woman who gave Natalie her wonderful grey eyes. Natalie turns to her mother. Which is to say, she looks into her mother's eyes; looks to see where Anna's eyes are directed. She follows Anna's gaze. Anna is looking at me. So Natalie turns back to me, smiles, and reaches out.

We are playing with her. We are getting reactions and for us, these

reactions make a human kind of sense. But for Natalie, our game is
pure geometry. Things within reach and things out of reach. Limbs
draw lines through space; eyes follow the lines. For Natalie, this is
not a game about people; it is a game about space.

There is a lot for her to learn. Between a game of lines and a game
of minds there is a great mountain of knowledge to be scaled, and
if you look into her eyes you can see her, climbing furiously. She is
emerging, bit by bit, month after month, out of the mathematics of
rote existence and into the heady, complex spaces of language,
symbolic logic, and, at last, an understanding of others. In the beginning my daughter's eyes were blank: windows on to an unlit room.
These days, they shine. Each day, they react and express more. Most
of the time, they smile; when she cries, they shine with a terrible
grey light.

This is a book about the nature of the eye. It is about all the eyes
that are, and ever have been, and may yet be. It is about how human
eyes see the world, and how other eyes see it. It is about what happens
to the world when it is looked at, and about what happens to us
when we look at each other. It is about evolution, chemistry, optics,
colour, psychology, anthropology, and consciousness. It is about what
we know, and it is about how we came to know it. So this is also a
book about personal ambition, folly, failure, confusion, and language.

This has not been done before. There are books of zoology, books
about evolution, books of vision theory, and psychology, optics, medicine, and biochemistry. There are atlases, and textbooks, privately
published labours of love, papers, bulletin boards, polemics. The
knowledge is there. The language in which it is expressed is sometimes
exquisite. But no one, to date, has ever attempted to reach into each
and every one of these territories to account for the eye. No one has
put the eye at the centre of a sprawling and epic story.

I soon found out why. Like Laurence Sterne's hapless autobiogra-
pher Tristram Shandy – whose life is lived more quickly than it can
be written down – I watched with growing horror as the pile of
unread research material grew, month after month, to dwarf the
material I had already mastered. My own daughter's eyes are an
admonition: I have watched as she has acquired colour vision,

empathy, complex attention, simple ball skills and even reading – all in the time it has taken me to barely scratch the surface of these subjects. And of course I had to range far beyond these human matters to fulfil my brief – to tackle other eyes, and other ways of seeing, and ask how all these eyes came to be, and wonder what they might become.

How to organise such diverse material? Two solutions presented themselves – as obvious as they were useless. I could describe how eyes came to be, tracing their evolution and development from their biochemical origins to their modern forms. And how dull that would have been: to plunge my reader-victims first into the toils of organic chemistry, then into the niceties of evolutionary interpretation, without a word said about why the story of the eye was at all worthwhile or interesting. This would hardly be a story at all, because it would have to fan out, as the eye has fanned out over evolutionary time, into myriad forms, countless small tales of variation and innovation. It would have resembled a failed experiment in hypertext fiction – lots of promise and great openings, but no satisfaction, no ending, no shape.

The obvious (and equally wrong) alternative was to write a ruthlessly simplified history of how certain men and women came to understand the eye. The trouble is, the eye is not one story. It is many. To preserve the chronology, every paragraph of the book would have had to begin with 'Meanwhile . . .' or 'Coincidentally . . .'

My solution is, I concede, a little strange. It is a reflection of my own journey through the literature of the eye – a tale of wonder, and confusion, and flashes of understanding.

Chapter One stakes out the territory. Can a single account, however casual and anecdotal, really hold such a vast subject together? Why would we want to relate the four-eyed world of the fish *Anableps anableps* to the meaningful glances of the higher primates; or the microscopic eye-spots of algae to the hallucinations artificially generated in the brains of human volunteers by the fields of a trans-cranial magnetic stimulator? The answer is in the chapter title: vision belongs to a 'commonwealth of the senses'. It is not a unique and privileged sense, but an exquisite manifestation of a drive common to all living things – vegetable as well as animal – to harness light. In Chapters Two to

Five, I trace the evolution of that commonwealth of the senses, and describe how different eyes came to fulfil different survival roles for different animals. The story is told episodically, from the point of view of the men and women who uncovered the eye's long history, and ends with an account of the oddities of our own human eyes.

Could it be that thinking arises as a response to seeing? Without eyes, would minds exist at all? Since written records began, philosophers have wrestled to equate the mechanics of the eye with the experience of vision. As our understanding of the eye has deepened, so their writings have shaped our investigations, helping and hindering in equal measure. Chapter Six looks at our theories of vision – again, through the lives and words of the people doing the theorising – while the rest of the book, except the last chapter, looks at how the problematic partnership of theory and discovery has tackled one of the more intractable problems of vision: understanding the retina.

Chapter Ten brings us up to date, by considering the future of vision. What does it mean to build an eye? At what point do we say that a machine genuinely sees? What extra senses might we give our own eyes and one day, will our eyes do more than see?

If this mix of history, science and anecdote has resulted in a mongrel of a book – well, so much the better. It is not meant to be a textbook, nor a history in the usual sense (a quick glance at the index will reveal, to the informed eye, some truly outrageous omissions). I have tried to give the reader a glimpse of what vision feels like as a subject of study and wonder; to give the artists and scientists, merchants and priests, adventurers and dilettantes who stumbled upon, and over, and through these mysteries, a voice. Wherever possible, I have let them speak in their own words.

I have dedicated this book to my daughter. Her eyes inspired it, and I hope that when she is older, she will glance through it, and tell me what she thinks of it. By then, she and I will be looking at each other through eyes that perform equally well.

But very soon after, the tables will turn.

The fat in the orbits of my eyes will shrink. The whites of my eyes will yellow. Probably my irises will acquire a curious, milky blue ring, an *arcus senilis* made of cholesterol – nothing to worry about, not a sign of any imbalance, 'just an age thing'. My brown irises will bleach, and if I live long enough they may even return to the baby-blue of my infancy.

Already, the gel that filled the back of my eyes at birth is partly liquid now – a process that began when I was three. With its delicate, ill-understood structure melted quite away – '*vitreous* humour' indeed! – the little flecks of collagen gathering inside my eyes are free to float hither and thither, like tiny spiders crawling across the panes of my pupils, a nuisance that will never go away.

If I reach seventy, there is an even chance that my lenses, exposed to light for so many years, will grow milky with cataract; a one-in-twenty chance that my retinal pigment epithelium, the retina's by then exhausted gardener, will fail; that blood vessels – seeking, in their clumsy way, to repair the damage – will worm their way into the retina, throwing everything awry, distorting what I see, and robbing me of sight entirely, from the centre out. (This macular degeneration is what I really fear. It is the largest cause of blindness for people over fifty-five; I used to smoke, and that makes me especially prone.) And there is a one in forty chance that my eyes, held in shape by the pressure of the liquid inside them, will clog and fill and swell, damaging my optic nerve; glaucoma: a treatable condition but often, in the elderly, relentless.

Maybe none of this will happen. (There will probably not be time: neither side of my family is particularly long-lived, and there is no reason why I should be an exception.) How well will I see my daughter, when she is the age I am now? The answer is worth remembering, a ray of light to pierce this gloomy talk of age and illness.

Age withers us. But our eyes – so delicate, so fragile – resist. Even as the rods in my retinas start to die, nerves will regroup to make the most of their failing information source. At eighty, my eyes will still see; if not with an adult's clarity, then at least with a child's more blocky, pixellated vision.

If I live to a good age, I will see my daughter for the last time as well as a three-year-old sees – as well as my daughter sees me now.[8]

ONE

The commonwealth of the senses

. . . .

Helen Adams Keller, daughter of a former captain in the Confederate Army, was born in 1880 in Tuscumbia, a small town in north-west Alabama. At nineteen months old, she came down with a fever and fell into a coma. When she awoke, she was blind. Her hearing deteriorated soon afterwards. What her illness was no one knows for sure, though meningitis seems likely.

Thanks to her own efforts and the extraordinary dedication of her tutor, Anne Sullivan, herself blind, Helen Keller went to college, wrote about a dozen books, met a dozen US presidents and travelled the world. She also found time to develop a lively appreciation of the plastic arts. 'I sometimes wonder if the hand is not more sensitive to the beauties of sculpture than the eye,' she wrote. 'I should think the wonderful rhythmical flow of lines and curves could be more subtly felt than seen. Be this as it may, I know that I can feel the heart-throbs of the ancient Greeks in their marble gods and goddesses.'[1]

In what sense was Helen Keller not seeing the statue?

If you ignore a wasp, it won't sting you. I know this sounds like an old wives' tale, but for Natalie's aunt Florence – a precocious ten-year-old – hosting wasps has become something of a party trick. Come the summer they gather around her in swarms. I've seen them walk across her eyelids. She sits very still, and she never gets stung.

Wasps have evolved a coarse-grained vision that enables them to map food sources on the wing. The curving routes they take past the picnic table are fly-bys: they're mapping where the lunch is. The downside of this curious style of vision is that the wasp cannot detect relative movement. As far as the wasp is concerned, everything is stationary. Move, and you confuse it. Run about screaming, waving a rolled up newspaper, and you vanish. To say this puts the wasp at a disadvantage is putting it mildly: why else do you think it evolved a sting?

The wasp cannot see things. It can only map them. So in what sense does the wasp have 'eyes'?

This chapter is entitled 'The commonwealth of senses' to convey two different but parallel ideas.

First, the eye is not a lonely miracle. It is a sense organ, similar in application to an ear, or a nose. Of course there are differences between seeing and hearing and smelling, but there are common features too, and these are important. If we look at vision as one of a family of senses, we can more easily understand where the eye came from, how it evolved, and how it works. This idea is explored in the first part of this chapter.

Next, I want to show that eyes are themselves a commonwealth: not one thing, but many, and supporting many different ways of seeing.

Human vision is a suite of abilities. Occasionally, following an accident or illness, one of these abilities fails. Moving objects cease to move smoothly; they disappear and reappear, apparently at random. Friends and relations become unrecognisable, replaced by impostors; everyday objects become objects of profound mystery. All the colour drains out of the world.

From the range of visual abilities we take for granted, many animals get by quite happily and successfully with just one or two. There are animals with light-meter eyes, sextant eyes, range-finding eyes, and eyes which exist solely to detect movement. The visual limitations of these animals never show up in the normal course of events, but only following some drastic change in their environment. A captive toad, though its gaoler thoughtfully provides it with dead flies, will starve to death waiting for a live fly to zip past its eyes. The native birds of Hawaii, lacking natural predators, ignore fast-moving objects; year by year, they are being transformed into brightly-coloured road kill.

Rather than just conduct a 'magical mystery tour' of different sorts of eyes, I have organised the second part of this chapter according to how they handle light. All eyes have to handle light, and light obeys the same laws everywhere. This means that even the most unlikely eyes will share some common features.

One of the less obvious, but extremely important, common features of eyes is the way they move, either by themselves or as part of a moveable head or body. This deserves, and gets, its own section in this chapter.

I end with a section that boasts the grand (if not grandiloquent) title 'The "problem" of consciousness'. We are aware of what we see; we appreciate the view. The humble mussel, on the other hand, has eyes that are scarcely more sophisticated than the motion detectors that turn on porch lights. It has neither the eyes nor the *nous* to appreciate the view. What about cats? Do they 'appreciate the view'? Are they conscious of what they see? Are they 'conscious' – in the usual sense of the word – at all?

This is not a book about consciousness (there are quite enough of those already). But, very briefly, I try to explain why I think it is legitimate to write about how other animals see and experience their world, and to do so without the annoying scare-marks – inverted commas and the rest – that litter more circumspect accounts.

1 – Seeing and touching

The eye owes its existence to the light. Out of indifferent animal organs the light produces an organ to correspond to itself . . .

J W von Goethe

Imagine that Helen Keller were restored to us. She and I are wandering around a gallery of statues. We are having a friendly competition. I am going out of my way to prove that sight is the monarch of the senses, quite different in kind and impossible to emulate or even describe. Keller, on the contrary, is out to prove that sight is not so special; just one of a family of closely related senses, and that one sense can stand in, quite easily, for another.

One handsome sculpture attracts our attention. (Perhaps it is by the seventeenth century sculptor, Jean Gonnelli. Born in Tuscany, Gonnelli went blind at twenty, long before he executed his best work. He is reputed to have modelled his portrait of Pope Urban VIII using touch.)

When I look at the sculpture, I sense the presence of a three dimensional object. When Keller touches it, so does she. I see these three dimensions by processing the subtle inconsistencies between the two-dimensional image projected on to my left retina and the two-dimensional image projected on to my right retina. I also make use of my familiarity with the way shadows work, and the way light behaves when it strikes an object. Keller experiences its three-dimensionality through touch, and correlates these sensations through her sense of how her fingers are oriented with respect to each other ('proprioception').

You might say that from where I stand, I see the whole scultpure in one go – a single glance – whilst Keller senses the object only where it touches her skin. But you could just as easily reverse the emphasis, and point out that Keller can curl her fingers around the object, whereas, to perceive it from different angles, I have to move my body.

And while I might like to think that I can take in an object 'at a glance', in reality my eyes are never still. Every third of a second they jolt or 'saccade', moving my gaze from one part of the object to another. My 'single glance' is a multitude of little fixations, not unlike the twitching of an insect's antennae, or a mouse's whiskers – or Keller's busy fingers, come to that. (A classic experiment in eye tracking by the Russian psychologist Alfred Yarbus reveals how we run our eyes over a face.[2])

I look at this object from a distance: to touch it, Keller stands within reach of it. If I step back, I can still see the statue. If Keller steps back, she loses sensory contact.

True enough, but is this really a qualitative difference between sight and touch? Each time I step back, I see the statue in less detail. Because my eyes are designed to find and exaggerate edges and lines

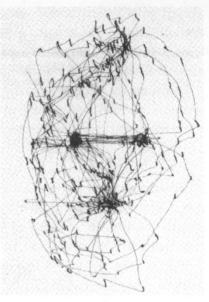

Alfred Yarbus's record of eye movements while
studying a picture of a girl's face.

– so that objects stand out from the mote and blur of the world –
this loss of detail is hard to appreciate. (Look up at the moon, and
its features appear razor-sharp: impossible to infer from this view
that these edges are actually wildernesses of shattered regolith.) The
further away I move, the less I see.

When Keller lets go of the statue, her sense of it is entirely lost,
instantly. Nothing passes from the object to her hands, the way light
passes from the object to my eyes. But what if it did? We can easily
imagine a technology that transmits tactile information about an
object to a pair of mechanical gloves. This has long been a commonplace of science fiction, and there are plenty of (admittedly, fairly
ungainly) examples on the market. So, this difference between touch
and sight is pretty trivial: merely a question of when a sense can or
cannot be used. Technology has been moving these frontiers about
for a very long time. (Excuse me while I polish my spectacles.) My
leading Keller out of reach of the statue is, philosophically speaking,
no more significant than Keller turning out the lights. Just because
I'm plunged in the dark doesn't mean I cannot see.

Are there any differences left?

Certainly. Keller cannot see colours – I can. This is a real and profound difference, as is my inability to tell, just from looking, what temperature the sculpture is, what consistency, how heavy it is, or how malleable.

Of course there are differences between touch and sight. But, Keller counters, the differences are not the point. It's the similarities that are important – the degree to which my eyes and her skin enable us to experience the same things in similar ways.

And she is right.

Since the early 1970s, Paul Bach-y-Rita has been building prosthetic eyes for the blind: not false eyes, not glass eyes, but fully working organs. Bach-y-Rita – a specialist in rehabilitation medicine at the University of Wisconsin-Madison – has helped the blind to see.

His eyes do not look like eyes. The earliest models looked like clothing. Bach-y-Rita's vests are worn either across the stomach or across the back. Sewn into the material are 256 mechanical vibrators (nicknamed 'tactors' because, when they're activated, the subject can feel their touch). A computer worn at the hip receives pixellated images from an ultra-low resolution video camera, worn on a pair of spectacles, and translates these images into mechanical vibrations, via the tactors. The upshot is a kind of Braille, or Pin-Art, vision.

Bach-y-Rita's subjects reported that after a couple of hours they were no longer aware of the tingling sensations generated by the tactors. They were able to navigate between obstacles, and, eventually, to recognise faces. When the 'view' changed – because they moved, or because something before them moved, they reacted appropriately. If someone screwed up a piece of paper and threw it at them, they would duck.

Even more suggestive is an experiment, reported by Daniel Dennett, in which a researcher, without warning, manipulated a zoom button on a volunteer's camera, making it seem as though he were hurtling forward. The volunteer raised his hands to protect his face. *But the volunteer's vest was strapped to his back.*[3]

The artificial sense Paul Bach-y-Rita bestowed upon his blind volunteers not only works like vision – it feels like vision. It seems that the mind is not overly fussy where it gets its sensory information from. What matters is what 'shape' the information takes. If visual information is received through the skin of your back, it only takes your brain a couple of hours to start seeing through your back. If your back starts itching, you won't mistake the itch for a flash of light. The 'shape' of an itch is different to the 'shape' of a face, and the brain knows how to deal with each.

The senses have become specialised over evolutionary time, but they are never entirely compartmentalised. Consider these photographs, taken by Jerome J Wolken of Carnegie Mellon University in Pennsylvania. They capture what happens when a rod from a frog's eye detects light. Rods come in two parts: a fairly normal-looking cell body, and a column made up of thousands of discs containing a pigment, rhodopsin. When exposed to white light, the pigment column expands, like a slinky, to twice its length, with no increase in width. In the dark, it contracts back. The rods behave just like muscle cells – and for good reason. In many functional respects, they *are* muscle cells. A muscle fibre expands and contracts in response to electrical stimulation. The retinal rod, too, responds to an electrical signal; but one that comes, not from a nerve, but from a biochemical reaction to light. Imagine what a working retina must look like on a cellular scale; a vast automated Pin-Art machine,

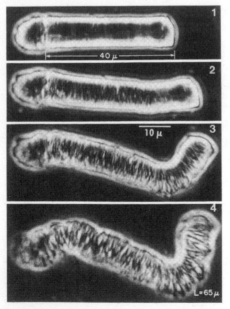

The mechanics of vision: a rod cell from a frog's retina expands when exposed to the light.

turning light to texture. At this scale, we can easily see the functional similarities between the eye and one of Paul Bach-y-Rita's vests.

If you swim in the sea, sooner or later you're going to end up looking like a fish. There are only a handful of really good ideas in nature and, whatever their genetic past, living things tend to evolve into one of just those few, very efficient, forms. This is *convergent evolution*.

If you hunt, you're going to develop eyes. There are very few really good forms

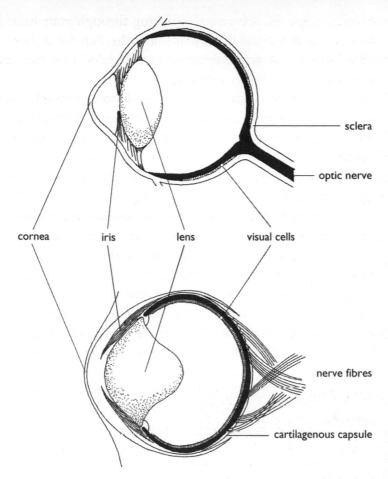

Convergent evolution: though unrelated, the vertebrate eye (*top*)
and the cephalopod eye have acquired many similar structures.

an eye can take, so we should not be surprised when quite unrelated
species evolve eyes that look remarkably alike. The eyes of cephalopods
(squids, octopuses and the like) and the eyes of humans are entirely
unrelated. Human eyes develop from specialised patches of skin;
cephalopod eyes are outgrowths of nervous tissue. Human retinas
look inwards, pointing their photoreceptors towards the back of the
eye; cephalopod retinas look out, pointing their photoreceptors
towards the light. Yet in striking ways, human eyes and cephalopod
eyes are virtually identical. Both work like cameras: a lens, behind
an aperture, focuses inverted images on to a photosensitive layer at

the back of the eye. Both have irises to control the depth of field, and both have lids to shield them from too much light.

Like Paul Bach-y-Rita's seeing-blind volunteers, the star-nosed mole *Condylura cristata* 'sees' through touch. Its nose has evolved into a mobile fleshy organ about a centimetre across. It is exquisitely sensitive, with five times as many nerves as run through the human hand. Its 25,000 'tactors' (to stretch a metaphor) are arranged across twenty-two 'fingers', so that the mole's nose is most sensitive at its centre. The distribution of nerve endings bears a more than passing resemblance to the retina of a mammalian eye.[4]

Condylura is not the only burrowing animal to have invented a touch-based 'vision'. Kenneth Catania, the man who soaked his feet in the bogs of North America to uncover the habits of the star-nosed mole, has also studied the naked mole rat *Heterocephalus glaber*, possibly the world's least prepossessing mammal, and certainly the closest anything warm and furry has ever come to impersonating the termite. The naked mole rat's looks are as ghastly as its habits – in place of an insect's antennae, it has grown its teeth forward like bayonets. But they are not just weapons; Catania's studies suggest the blind mole rat's teeth are as sensitive as a star-nosed mole's nose.

Sight is one of those few really good ideas abroad in the world, and an ability to see is something that many very different animals acquire.

2 – Handling light

How many different kinds of eye are there? Most eyes fall into certain broad categories: there are single-chambered eyes like ours; insects and crustacea have complex eyes, whilst many animals get by with very primitive eyes that lack even the rudiments of lenses. Outside these broad categories there are numerous oddities: the tiny (3 mm-long) crustacean *Copilia* scans the world line by line, its retinas in constant motion.[5] A close relative of the starfish, the brittlestar *Ophiocoma wendtii*, is one huge complex eye, its whole surface punctured by little eyespots linked by nerve bundles running just under the skin.

The differences between these eyes can be dizzying, and yet all of them have evolved in response to the way light strikes our spinning, weather-racked planet. If we look behind their differences, we will uncover their common principles.

The following four sketches – in no particular order – describe the ways in which light sculpts the eye.

True eyes gather light. The largest eye ever found belonged to a giant squid, *Architeuthis dux*, a monster even by the standards of its own species, found alive and stranded by three fisherman in Timble Tickle Bay, Newfoundland, in 1878. Before they chopped it up for dogmeat, they measured it: its tentacles were ten and a half metres long, and its eye was forty centimetres in diameter – as wide as my computer screen. Giant squid are predators, moving vertically through the water in search of prey – mainly other squid, although they have been known to make a light snack of baby whales. (Adult sperm whales repay the compliment; giant squid form at least eighty per cent of their diet.) Like most ocean species, giant squid cruise at a preferred depth, a kilometre below the surface, the very point where sunlight gutters out, leaving the ocean in perpetual night. This is the reason for the giant squid's gigantic eyes: it needs them so it can see its tentacles in the dark. (They aren't muscular, and the giant squid has to dart forward to swallow the prey its tentacles have captured.)

The giant squid is a visual predator, operating at the margins of sunlight. Its big eye, with its correspondingly big pupil, gathers more light than would a small eye.

A kilometre below the 'cruising height' of the giant squid, even further from the reach of sunlight, lurks an even more daunting sea monster, *Mesonychoteuthis hamiltoni*, dubbed the 'colossal squid' in 2003 after a young specimen was caught in the Antarctic. No intact adult specimens have ever been recovered, and no eyes have survived sufficiently undamaged for close study. But the beaks of the elusive adults have turned up in the stomachs of sperm whales, and using them as a measure, Steve O'Shea, senior research fellow at Auckland University of Technology, predicts that a fully grown colossal squid would have a torso (or 'mantle') about four metres long. After his autopsy of the young *Mesonychoteuthis*, O'Shea wrote to one correspondent: 'I can assure you that the eyes of *Mesonychoteuthis* are larger than anything known for [the giant squid] *Architeuthis*.'[6]

The eyes of abyssal monsters like *Mesonychoteuthis* and *Architeuthis* gather all the light they can, but animals exposed to ordinary sunlight have carefully to regulate their sensitivity, and shield themselves from too much light. Daylight on Earth is a million times stronger than dim starlight, and night rapidly follows day. Eyes must not only feed on light; they must respond to extremes of glut and starvation.

In nocturnal animals like owls and opossums, the opening of the eye – the pupil – is almost as wide as the eye itself, and any exposure to direct sunlight would be utterly blinding. To protect their eyes from damage, and to adapt quickly to changes in light level, many eyes have evolved pupils that can shut out the glare at an instant's notice. A cat's iris is slit vertically, and geckos have slit pupils that are notched to create a series of pinholes: both muscular arrangements can respond more quickly and more dramatically than can the human iris. Our pupils, evolved for daylight living, dilate to four times their constricted width when the light grows dim, granting a sixteen-fold increase in their light-gathering powers.

Find a friend, and stretch a rope out taut between you. Give the rope a flick. A wave will travel down its length.

That's light.

Flick your end of the rope up and down, and the wave will travel vertically. Flick the rope from side to side, and the wave will travel horizontally. In the same way, a wave of light can travel in any plane.

In the vacuum of space, there is no plane along which light waves preferentially travel. However, when light hits the Earth – when it passes through a cloud, or reflects off an ocean – the plane along which a light wave is travelling can make a difference. Light travelling in one plane will go one way; light travelling in another plane will go another; light in yet another plane will be absorbed.

To picture how light is filtered in this way – or 'polarised' – find a picket fence. Stand on one side of it, stand your friend on the other, and stretch the rope through a gap in the pickets. Flick your end of the rope up and down; the wave will travel vertically along the rope, through the gap in the fence. Flick the rope from side to side; the wave will travel horizontally until it reaches the fence, where it hits the pickets and dies. Polarised light is made up of waves vibrating in one plane, and it is made by filtering out all the light waves travelling in other planes.

Every eye on earth detects light using rhodopsin, though this is sometimes supplemented by other pigments. A molecule of rhodopsin changes shape when it is hit by a wave of light. But there is a catch: the light has to hit the molecule at a certain angle – along a certain plane – for the shape change to occur. The pernicketiness of rhodopsin molecules makes no difference to human vision, as our seeing cells arrange their pigment molecules every which way. The likelihood is that a wave of light entering the human eye, on whatever plane, will hit a correctly orientated rhodopsin molecule eventually.

The compound eyes of insects and crustaceans, however, are effectively colonies of many hundreds or even thousands of tiny eyes called ommatidia. Conditions in a compound eye are invariably cramped, so each ommatidium stacks up its pigments in a neat space-saving column. This means that fully three-quarters of its pigment molecules lie on the same plane. This tidy arrangement means that the pigments react strongly to light travelling in one plane, and hardly at all to light travelling in another. On first acquaintance, this design seems disadvantageous. What if the insect stumbled into an area lit by the 'wrong' sort of polarised light? It would be rendered virtually blind.

Happily, such dramatic effects do not occur in nature. What does happen is that the earth's atmosphere itself acts as a weak polarising filter. This is useful: even on an overcast day, animals sensitive to polarised light can detect a column of light, passing directly from the sun to their eyes. They know where the sun is in the sky, at every moment, without having to look directly at it. Bees, ants and many other animals use their polarised vision to navigate: they steer by the sun. (Nocturnal insects like the dung beetle, not to be outdone, steer by moonlight.[7])

Water, and other non-metallic reflecting surfaces, also polarise light. At a certain angle, all light reflected off water is polarised on a plane parallel to the water's surface; some water bugs, including the water boatman *Notonecta*, use this light to spot new water sources. Underwater light is even more strongly polarised, and many aquatic animals are aware of this. Polarised light reflecting off an object will have a characteristic pattern; using these patterns to identify prey makes a great deal of sense in the ocean, where the colours of things shift markedly with small changes in depth. Even transparent animals like jellyfish cast polarisation 'shadows'.

Some of the cephalopods – octopuses, squid, cuttlefish and the like – may use polarised light to communicate. By subtly ruffling the tiny platelets that make patches of their skin iridescent, cuttlefish can not only alter the colour of their skin, but also its polarisation pattern, providing them with a secret communications channel, invisible to predators.[8]

Although polarisation vision has been harnessed in many different ways, in most cases the evolutionary pressure to adapt to polarised light may run the other way, towards eyes that are blind to polarised light. Polarisation can be confusing. The polarisation of reflected light, for example from a leaf, depends on the position of the sun, the angle of the leaf, even upon whether it has rained or not. Insects that use polarised light for navigation generally restrict this vision to the upward-looking portion of their eyes. Elsewhere, the *ommatidia* (the elements, corresponding to a small simple eye, that make up the compound eye of an insect) are twisted along their length, so that each layer of photopigment is pointed in a slightly different direction to the layers above and below. The honeybee relies on polarised light to navigate, but most of its 5,500 ommatidia are twisted through 180 degrees, and are thus robbed of any sensitivity to polarised light.

Although it is sometimes said that humans have no sense of the polarisation of light, this ignores the fact that we are a tool-making species. For several centuries humans have made tools that show us how light is polarised.

Raudulf's Saga, a short family tale preserved in an early fourteenth-century manuscript, describes the visit of the Norwegian king Ólafur (Saint Olaf, died 1030) to a rich and wise farmer, Rauðúlfur. Late one night, one of Raudulf's sons, Sigurður, boasts that he can 'discern the motion of heavenly bodies . . . although I do not see

[them]', a neat trick which enables him to 'discern all hours both day and night'.

'This is a great skill,' says King Ólafur, and he's not kidding. In the morning, he puts the boy's talent to the test. 'The weather was thick and snowy, as Sigurður had predicted . . . The King made people look out and they could nowhere see a clear sky. Then he asked Sigurður where the sun was. Sigurður gave an unambiguous reply.'

Thus far, we are comfortably in the land of the tall, if not the fairy, tale. Sigurður's supernatural talents are too good to be true. What happens next, though, must give us pause, because it turns out the King has a sure-fire way of checking whether Sigurður is right.

'Then the king made them fetch the sunstone, and held it up and saw where light radiated from the stone and thus directly verified Sigurður's prediction.'

The account could not be more matter-of-fact. The king's 'solar stone' is no fantasy. The storyteller presents it, without fanfare or fuss, as just another piece of everyday royal kit; as familiar as a watch, as utilitarian as a compass.[9]

The king's sunstone may have been a piece of the mineral cordierite; pebbles of the stuff are scattered along Norway's coast. Using a neatly cleaved crystal of cordierite, it is easy to tell the direction of light polarization: the stone's colour will change from blue to light yellow when it is pointed towards the sun. Or the stone may have been Iceland Spar, an optically pure form of calcite. How wonderful if it were: the eyes of trilobites – the earliest eyes we know of, five hundred and fifty million years old – have lenses made of the same stuff.

The sun emits energy at different wavelengths, and visible light is just a tiny part of the solar spectrum. Reading across that spectrum, from long wavelengths to short ones, we begin among low-energy, long-wavelength radio waves. Passing through the infra-red, we come upon low-energy visible red light. Once through the colours of the visible spectrum, we watch as high-energy visible blue light fades into the ultraviolet, and this itself quickly gives way to short-wavelength, high-energy gamma rays.

Visible light is the part of the electro-magnetic spectrum that penetrates most easily to the planet's surface. Most of the energy we receive from the sun arrives as visible light, peaking at around 500 nanometres (0.0002 inches), the wavelength of blue-green light.

Every animal sees a slightly different spectrum, depending on what use it makes of particular wavelengths of light. At the long-wavelength end, some fish and butterflies see a little into the infra-red, which gives them extra visual sensation at dawn and dusk. But the infra-red band is hard for the eye to harness, and no eye can see very far into it. Infra-red is absorbed by the earth's atmosphere, water vapour, and the water surrounding living cells. Infra-red wavelengths are long, which accentuates their wave-like properties. They tend to bend (or diffract) around corners, like waves around a sea wall. Images made in infra-red light are blurred; load infra-red film into your camera, and you will find that a human face will always come out as a white blank: a hot surface, entirely lacking in definition. This is why animals that use the infra-red spectrum use it for target-finding and navigation, rather than the examination or manipulation of objects. The infra-red-sensing pits near a snake's eye enable it to locate its prey, while some beetles navigate towards food-rich forest fires by detecting the infra-red energy of the blaze.

Humans can see a sufficiently powerful infra-red source, as the vision pioneer George Wald discovered while developing infra-red viewing devices for the US Army Board of Engineers during the Second World War. (The devices were not particularly sensitive, so they built prototype infra-red searchlights, so powerful they triggered a dim red glow in the eye and a distinct and worrying warming of the skin.[10]) In the real world, however, infra-red vision has little to offer the human species, a primate whose main visual behaviours require precise, focused vision.

Insects, birds, fish and mice see shorter wavelengths than humans, well into the ultraviolet. Many flowers boast striking patterns, only visible in ultraviolet light, to attract pollinating insects. Even with their superb visual acuity and excellent colour sense, extending well into the ultraviolet, kestrels find it hard to spot the drab voles which are their favourite food. Happily for the kestrels, however, voles communicate by leaving trails of urine – indeed, they pee almost continuously – and vole urine reflects ultraviolet light. For kestrels, hunting for voles is simply a matter of following the arrows.

Ultraviolet light, like infra-red light, is largely blocked by the earth's atmosphere. But it is extremely energetic – as anyone who forgets their sun cream on the beach can testify. It is energetic enough to trigger photoreceptor cells sensitive to far longer wavelengths. The challenge for the ultraviolet-adapted eye is to develop a system that doesn't simply fire randomly and everywhere the moment sunshine hits it.

No big animal sees far into the ultraviolet. The larger the eye, the more light it can take in, and there must be a point at which the potential damage ultraviolet light can do when focused outweighs its usefulness. Many birds and insects have evolved to see ultraviolet wavelengths, but they live for only a short time, dying before the damage becomes significant. Large mammals have much longer lives, and their exposure over years could destroy their eyes' photopigments and turn their lenses cloudy.

Mammals that are active at dawn and dusk require less shielding against ultraviolet light, and benefit more from good vision in blue and near ultraviolet light. (Any rider will tell you that horses can pick their way through the dusk with an ease that borders on the uncanny.) Animals which are active in daylight have evolved many ingenious systems for shielding their eyes against ultraviolet light. Squirrel eyes have yellow-tinted lenses, while the photoreceptors of fish, birds and turtles are tipped with coloured oil droplets, which may also play a role in their perception of colour. The particular, and unique defence evolved by primates, including human beings, involves a yellow pigment that absorbs ultraviolet light almost completely. This 'macular pigment' covers the whole fovea, which is why it turns a bright lemon-yellow when the retina is exposed to the air – an effect which first startled the physician and inventor Samuel Soemmering in 1795.

By far the most effective ultraviolet filter in the human eye is the lens. Our lenses are extremely efficient filters of ultraviolet light, reflecting our daylight habits, but with the lenses removed, human eyes can perceive ultraviolet wavelengths, something which, though barely wondered over, must have been apparent to ophthalmic patients for at least two thousand years. In the first century AD, Roman doctors routinely displaced and removed irreversibly swollen and clouded lenses from the eyes of their patients. The condition they were treating, cataract, is still with us, and still irreversible. Since 1947, it has been possible to replace the lens with a plastic substitute. Before artificial lenses were available, however, those who had their lenses removed by surgery found that they could see into the ultra-violet; blues were clearer and richer, and ultraviolet light, energetically triggering every photoreceptor it hit, was a blueish-white wash.

Because most wavelengths are visible to most animals, we can be forgiven for thinking that most animals see the same colours, but many animals have very poor colour vision. Colour is just a way of distinguishing objects from each other. Where daylight is more or less the same colour all the time, objects retain their native colours, and the need for complex colour perception is quite low. In 2000, the Taiwanese neurophysiologist Chuan-Chin Chiao showed that daylight, filtered through a forest canopy, changes colour only very slightly.[11] No wonder that mammals, with very few exceptions, make do quite happily with 'dichromatic' vision: in human terms, they are colour-blind.

For some forest-dwelling foragers, impoverished colour vision may confer a survival advantage, since – for reasons I shall explore in Chapter Nine – it goes hand in hand with a greater sensitivity to lustre. A camouflaged object in dappled light is immediately noticed by a colour-blind person: its distinctive sheen gives it away. Marmosets may be natural camouflage-breakers; the females enjoy three-colour vision quite similar to our own, while the males are born with two-colour vision – and it is the males who are better at spotting food sources.[12]

In environments where the colour of daylight fluctuates a great deal, colours vary hugely, and several dimensions of colour vision are a real asset. The colourworld of reef fish is one such, very different from our own, where the bright blues and yellows of many reef fish function as camouflage.[13] Water absorbs sunlight, robbing it one by one of wavelengths, the deeper one goes. These changes are quick and dramatic: reds, oranges and yellows vanish from human percep-tion within about ten metres. For the same reason, the 'blueing' of objects with distance is far more marked underwater than on land. To keep tabs on the coloration of their surroundings, some reef fish have up to twice as many different colour receptors in the eye as do humans.

But the record for the most dimensions of colour vision is held, not by reef fish, but by brightly-coloured stomatopoda or mantis shrimps. Mantis shrimps are not shrimps. Some are thirty centimetres long. They will attack and attempt to eat just about anything, and have been known to break a diver's fingers. A blow from the claws and head of some species is almost as powerful as that from a .22 calibre bullet. In 1998, a green mantis shrimp put its head through the half-centimetre thick glass of its tank at the Yarmouth Sea-life Centre.[14] Look into its eyes, and there are more surprises still. Some species of mantis shrimp boast *sixteen* kinds of narrow-band colour detectors. Behavioural experiments have shown that these provide them with true colour perception. The evolutionary pressure on the mantis shrimp to develop such sophisticated visual apparatus appears to have been two-fold. It operates in the shallows, where colours vary drastically with small changes in depth, and it relies on displays and dances to govern its behaviour with other mantis shrimps.[15] A colour-blind mantis shrimp will not be able to side-step the aggressive approaches of rivals; even more seriously, it will not be able to locate and engage a mate.

Light reveals nothing about how far it has travelled. A ray of light arriving from a distant star makes no more fanfare on its arrival than does the glint reflected off the rim of a cup. How, then, are we to perceive depth?

William Molyneux, the Anglo-Irish politician and physicist, put his finger on the central difficulty in his work, *Dioptrika Nova*. Published in 1692, this was – contrary to the impression given by the title – the first book on optics written in English. 'For distance of itself,' he writes, 'is not to be perceived; for 'tis a line (or length) presented to our eye with its end toward us, which must therefore be only a point . . .' To put it another way, we only ever see light when it hits us. We cannot stand to the side of light and measure the distance it has travelled. So how can we tell, just by looking, how far away things are?

To measure the distance of a near object, we might use muscular

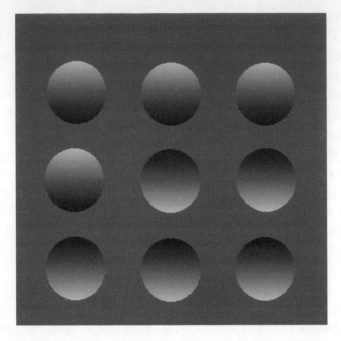

information from the eyeballs, as the eyes converge slightly to focus on a near object. This seemed like a plausible explanation for how we judge distances of up to six metres. But even if this mechanism existed (it doesn't), this still left the problem of how we judge the distance of faraway objects. This had to involve a different mechanism, because when we look at objects further than about six metres away, both eyes are looking straight ahead. But there seemed to be no difficulty in explaining our depth perception. Since the Renaissance, painters had demonstrated, time and time again, how well the eye infers three-dimensional forms from environmental clues like shading, perspective, and occlusion (the way things get in the way of each other). Some of these lessons are part of folklore: the grass *is* always greener on the other side of the fence, because individual blades of grass are more clearly seen from the side than from above. A field of grass hides the earth better when looked at from the side.

David Rittenhouse (1732–96) was a clockmaker by trade and an astronomer by inclination. Self-educated, he built the first astronomical telescope on American soil, and with it in 1768 he discovered the atmosphere of Venus, independently of its earlier discovery by the Russian Mikhail Lomonosov in 1761. He also came up with the optical illusion above (charmingly dubbed the 'muffin pan' illusion by the vision researcher Donald Hoffman). Which circles are muffins, which are empty dishes? The answer depends on which way up you hold the book. Rittenhouse's illusion reveals how much our sense of three-dimensional form relies upon the way our environment behaves. We assume that everything is lit from above, and this is how we interpret the shaded circles.

Leonardo da Vinci wrote very well about this kind of thing. His subjects included shadows, the way the eye handles detail, and the way the colours of things shift blue-wards the further away they are. He also makes a stab at explaining stereopsis: the indefinable 'is-ness' three-dimensional objects acquire when looked at with two eyes:

> . . . a painting, though conducted with the greatest art and finished to the last perfection, both with regard to its contours, its lights, its shadows and its colours, can never show a *relievo* equal to that of the natural objects, unless these be viewed at a distance and with a single eye . . . because a painted figure intercepts all the space behind its apparent place, so as to preclude the eyes from the sight of every part of the imaginary ground behind it.[16]

This explanation is not quite sufficient. It assumes that every object is small enough, or near enough, to occlude slightly different parts of space in the view of each eye. This is by no means always the case. Neither can he explain why we see a three-dimensional figure whenever we fuse stereoscopic drawings.

The trouble with Leonardo's explanation is that it relies solely on optics. It took nearly two hundred years before someone realised that optics alone could not explain the effect. That someone was Charles Wheatstone (1802–1875), inventor of the concertina, inventor of the telegraph, and Victorian polymath. In 1838, Wheatstone subtly revised Leonardo's argument. Out of one eye, Wheatstone says, a good enough painting can be confused with a real scene. But look at it through two eyes, and the illusion is broken. Why? Because 'When the painting and the object are seen with both eyes, in the case of the painting two *similar* pictures are projected on the retinas; in the case of the solid object, the two pictures are *dissimilar . . .*'[17]

The stereoscopic drawings that head the chapters of this book are from Wheatstone's original paper. Each features a pair of dissimilar drawings of the same object. Go cross-eyed, to fuse the paired images into one, and you will see a single three-dimensional figure. 'I have employed only outline figures,' Wheatstone explained, 'for had either shading or colouring been introduced it might be supposed that the effect was wholly or in part due to these circumstances, whereas by leaving them out of consideration no room is left to doubt that the entire effect of relief is owing to the simultaneous perception of the two monocular projections, one on each retina.'

Soon afterwards, Wheatstone smartened up the demonstration equipment he had thrown together for his Royal Society talk of 1838, and invented the stereoscope, 'an instrument which will enable any person to observe all the phenomena in question with the greatest ease and certainty.'

Anyone old enough to remember looking through red plastic ViewMasters as a child knows intimately what a stereoscope can do. Two photographs of the same scene, taken five centimetres apart (the average distance between human eyes), are mounted in an optical device that helps the viewer fuse these images into one. The result is a single, captivating three-dimensional image. (Strictly speaking, you don't need a stereoscope to see the image; go cross-eyed until the photographs overlap, and the images will fuse by themselves.)

One curious by-product of the stereoscope is a niggling sensation of vertigo. Nobody has eyes *exactly* five centimetres apart. Slight differences in stereopsis between the viewer and the apparatus can exaggerate the 3-D effect of the photograph. One gets the (perfectly justified) feeling that one is looking out of somebody else's eyes.

Wheatstone followed with the pseudoscope which, by means of prisms or mirrors, transposed the optical images of an object or stereogram as they were normally brought to the eyes. It never caught on: plunging into a slightly alien three-dimensional world was one thing, but having it turned inside out was quite another.

Wheatstone's stereoscope demonstrates that when we fuse our left-eye and right-eye views of the world, the tiny discrepancies between

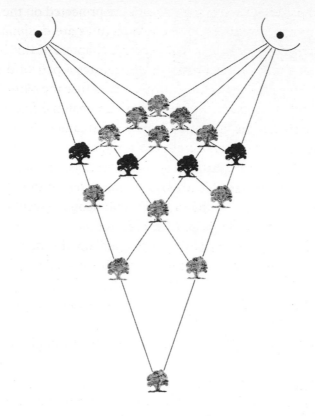

them give us a sense of depth. What he could not do at the time was explain how these tiny discrepancies are spotted, measured and interpreted by the eye and brain. How do we stitch these slightly dissimilar views together in reliable ways? When you walk into a thick and tangled forest, how do you know there are, say, four trees a few feet away from you, and not three trees, further away? Confronted by four trees in a row, and with no foreknowledge of how the images in one eye match up with the images in the other, you can make *sixteen* perfectly sensible assumptions about where the trees are. Twelve of them will be wrong. What if you were faced with ten trees? There are a hundred possible places the trees could be – that's *ninety* opportunities for error. Now think about how many trees there are in a wood. No wonder people thought – even as late as the mid-sixties – that environmental cues like shading, perspective and occlusion were the main engines of depth perception, and that stereopsis was a mere curio.

Bela Julesz (1928–2003) changed all that. In 1953, Julesz, a PhD

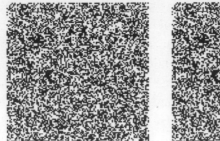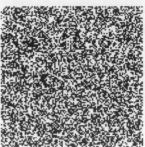

student at the Hungarian Academy of Science, embarked on a thesis analysing the encoding of different televised images. It was a labour of love – apart from anything else, television had yet to arrive in Hungary.

The thesis was never published: in 1956, Russian tanks drove into Hungary. Julesz, armed with a copy of Plato's *Apologia of Socrates* (his favourite book), fled with his wife from Hungary to the United States. He went to Bell Laboratories in New Jersey, where he worked as a radar engineer. Later he joined a group dedicated to 'reducing visual information in pictures without perceptual deterioration.' Julesz thought that some familiarity with human vision would be useful, and so he started to study psychology.

What he read startled him: apparently, depth perception was a nebulous and complicated process that ran the gamut of the brain's machinery, from memory to symbolic logic. This was news to Julesz who, as a former radar engineer, knew damn well that all you needed was two eyes: 'After all, in order to break camouflage in aerial recon-naissance, one would view aerial images (taken from two different positions) through a stereoscope, and the camouflaged target would jump out in vivid depth.'[18] Julesz realised that, to prove his point, he needed to show that even perfect camouflage will fail the stereoscope test. So he created the random dot stereogram. Take a look at Julesz' original 1959 figure. In both pictures, there is a simple shape lurking among the dots. In neither can you make out the figure, because it is perfectly camouflaged: a white figure, covered in random dots, against a white background, covered in random dots.

Go cross-eyed, fusing the squares into a single square. Now do you see the figure?

Random dot stereograms are easy to make. Although they have always been generated on computers (so no need for glue, scissors and obsessive neatness) the process is simple. Take two sheets of paper printed over with the same pattern of random dots, stack them together, and cut out a shape.[19] Take two more identical sheets, and lay one shape over each sheet. Adjust the shapes so that both lie on exactly the same spot on their sheets. Then displace one figure horizontally by a few millimetres. Go cross-eyed, fusing the pictures, and the shape will appear.

The left- and right-hand pictures are equivalent to left-eye and right-eye views of the same scene. Depending on which way you displaced the shape – towards or away from each other – the shape will either leap out at you, or appear as a hole in the image. Either way, the shape will reveal itself. As Julesz said, 'in real life, there is no ideal camouflage.'

If, confronted by ten trees, we can make ninety wrong guesses about their number and position, how can we possibly successfully fuse the hundreds of dots that make up a random dot stereogram? The answer is straightforward: trees are big, complex, gnarled objects. Every square inch is a mass of surfaces, shadows, occlusions – a mass of dots, in fact. Stereopsis works by processing the dots.

The views of both eyes are mapped over a single sheet of neurons in the visual cortex of the brain. When both views are laid over each other, a special class of neurons spots the inconsistencies: a spark in the left eye doesn't quite overlap with a spark in the right eye; a smear in the right eye doesn't extend quite as far as the same smear in the left eye. We know where the trees are, because the scene before us has *already* been sifted for all those tell-tale little disparities between the left and right eyes. We see where the trees are because we already know where to look.

An apology: some of you won't have been able to see anything in the stereogram. About one in twenty people will have little or no experience of anything discussed in the last few pages – because they are stereoblind. Many do not know about their handicap. Stereopsis is subtle, and in its absence, environmental cues *are* enough to see us through most everyday situations.

The conundrum of the observer in the forest, confused by a handful of trees, only holds while the observer remains absolutely still. The moment the observer moves, the precise location of the trees becomes easy to judge, thanks to another technique for depth perception, parallax. Rock your head from side to side, and the way objects seem to move relative to each other will give you extremely accurate information about their relative distances. Parallax is used throughout the animal kingdom. Grazing animals generally have eyes mounted on the sides of their heads, to give them all-round vision. Because their fields of view barely overlap, stereopsis plays no part in their depth perception. Instead, they use parallax to judge the approach of predators. Predators use parallax, too. The preying mantis rocks its head from side to side to judge the distance of its prey.

Compared to parallax, stereopsis is a relatively rare and unusual form of depth perception; as far as we can tell, only a handful of lucky primates enjoy it. In 1995, towards the end of a career which revolutionised our understanding of the early stages of vision, Julesz came up with a just-so story to explain why primates might have acquired this extra, super-subtle sense of depth. Since there is no camouflage in 3-D, stereopsis probably evolved first in our insectivorous lemur ancestors, enabling them to spot insects that freeze, blending into the foliage at the first sign of danger. In other words, the wonderful 'is-ness' of the three-dimensional world may be sensory overspill from a quite different primate adaptation: the ability to see through camouflage. This would certainly explain why some insects have embarked on a hugely expensive counter-counter-tactic, remaking their bodies as twigs and leaves.

3 – The moving eye

Vision itself is a dynamic process. There is little in the world that stands still, at least not as imaged in our retinas, for our eyes are always moving. The visual system is almost exclusively organized to detect change and motion.

Haldan K Hartline[20]

Our eyes provide us with a picture of the world. It is an odd, magical picture, better than any film or video. In a moving image, everything is set afloat upon the picture frame, and the camera must exert an iron discipline if things are to retain their proper relationships. Move the camera too quickly, and the world lurches. Zip the camera from object to object to object, and space crumples. The eye, however, is free to explore the space it inhabits. It flicks, glances, pans, tilts – all in, literally, the blink of an eye. Yet spatial relationships are preserved. Stationary objects always appear stationary, and moving objects always appear to move.

The picture provided by our eyes is magical in other ways. It highlights the things we are interested in, enhances them in countless ways, and always puts the object of the greatest interest in the centre of the frame. None the less, this picture is inferior to the pictures provided by a camera, since virtually none of it is in focus. And, unlike a camera, it is unreliable. A camera records what is really out there, but the eye doesn't. In the picture provided by our eyes, whole objects can fail to register. An entire event can come and go, and leave not a trace. A cynic might say that the magic of such a picture lies in the fact that we never really notice how bad it is.

The picture we recieve through our eyes is odd. It is magical.

Which is to say: *it is not a picture.*

Although we think of vision in terms of images, images are not the be-all and end-all of vision. The natural world is full of eyes that cannot generate images. Where, then, is the common ground between our eyes and others? Is the story of the eye one story, as I have claimed, or is it many?

One way to understand how all eyes are related is to put aside, for the moment, the different experiences of vision, and instead look at what eyes actually do.

All eyes move. Either they come with their own musculature, or they are mounted on moveable heads or bodies. The significance of eye movement was slow to dawn, but today it is the focus of the very latest researches into perception and attention. At first, only human eye movement seemed interesting. Now, as we learn about the eye movements of very different species, we are beginning to uncover certain universal principles, on which depend many different forms of vision.

*   *   *

For the staid and sedentary professors of the University of Prague, medical graduate Jan Purkinje (1787–1869) must have cut an unnerving figure. His rooms were furnished with whirligigs and swings. Regularly, he would spin himself into a nauseous stupor. Purkinje, born to a family of Czech peasants, began his adult life as a teacher in a religious order. This did not last long; the 'continuous slavery to the superiors whose lives and dignity did not always come up to my expectations' soon led him to shed his monk's habit; by his own account, he walked the three hundred kilometres to Prague, where he supported his philosophical studies by private tutoring.

Purkinje was interested in the scientific value of his own experience. He believed that the results he obtained from his own body were more useful than any he might obtain from animals, or from fatally ill patients. When he began to study the physiology of vision, he took a variety of drugs, in a series of risky experiments that were looked on askance by his peers. Quite what they thought of his other abiding fascination is not recorded: Purkinje was particularly

Jan Purkinje in spin.

interested in a new method of pacifying the insane which had come out of Britain in the 1810s, 'which is now become of so much consequence in the cure of maniacs of almost every description.' The best treatment for mania, apparently, was to spin the maniac into stupefied and wobbly submission.

'The debility arising from swinging is never to be dreaded,' declared Joseph Cox, the inventor of the spinning chair: 'it is generally accompanied by sleep and a sense of fatigue, while the slumbers thus procured differ as much from those induced by opiates, as the rest of the hardy sons of labour from that of the pampered intemperate debauchee.'[21] Cox had a point: at a time when sedation was a hit-and-miss affair, accomplished by all manner of means – vomiting, purging, bleeding, digitalis, bathing, blisters, camphor, sedatives, stimulants – rotation in a chair was certainly one way of calming people down.

Purkinje, who spent no little effort exploring the links between movement and vision, built an optical toy which screened looped animations; one of his 'movies' shows him taking his home-built rotating chair for a spin. He used it to induce vertigo, stretching the abilities of his eyes to reveal – in their exhaustion and failure – the reflex that stabilises vision.

Purkinje understood that, first and foremost, the eye moves to compensate for the body's movements. A moving eye keeps the world still. It does so by means of a reflex, which Purkinje described with admirable exactitude. Briefly, Purkinje's reflex – the 'vestibulo-ocular reflex' – is triggered whenever our body's movements disturb the delicate hairs growing in the liquid-filled canals of the inner ear. Nervous responses from the hairs tell us how we are moving in space, and this information, which helps us balance, also triggers involuntary movements of our eyes. Thus, stationary objects continue to appear stationary, even when we are moving. If, for example, I look at something while moving my head, the vestibulo-ocular reflex compensates by moving my eyes in the opposite direction, cancelling out the head movement.

Spinning sloshes the fluid about in our ears, upsetting our balance. This loss of balance in turn confuses our vestibulo-ocular reflex. We not only feel as though we are falling – it really looks as though we are falling, as the world zooms past our eyes; as if gravity had changed direction, pulling us, not downward, but to the side.

Purkinje's work was based on the idea that optical illusions reveal the underlying mechanisms of vision. Extending this insight to the world of the senses generally, he evolved a biological science that was not purely mechanical, but made room for feeling and – eventually, in the hands of his successors – mind. Purkinje's successes were won in the teeth of often savage opposition. Hostility was directed both at the man himself (as a Bohemian, his ethnic background was a big hindrance to him) and at his working methods. What were his peers to make of this supposedly serious scholar, stumbling about the corridors, his eyes bleared by belladonna (a cosmetic eyedrop historically favoured by young women because of the way it dilated and paralysed their pupils)? When he introduced demonstrations and practical laboratory work into his teaching, this was simply intolerable, and there were calls for him to be stripped of his professorship.

Purkinje survived largely thanks to a well-placed friend, the German poet and polymath Johann Wolfgang von Goethe (1749–1832). A gifted natural philosopher, Goethe had also attempted to formulate a science concerned with subjective experience, and wrote extensively on colour perception, with mixed results. With Goethe's backing, Purkinje was able to establish intellectual credentials for his new, physiological science. Much of his own work, however, was forgotten. For years after his death, he was better remembered for discovering the forensic potential of fingerprints, and for purifying blood plasma, than for his greatest scientific work: the anatomy of optical illusions. This work, which had appeared under such provocative titles as *Contributions to the knowledge of vision from a subjective perspective* and *New subjective reports about vision* had to be rediscovered, after his death, by a new generation of psychologists.

His shade haunts almost every chapter of this book.

Most animals care very little for the substance of things. They are much more interested in where things are going. If it moves, it matters. Moving targets may be threats, meals, or mates. The earliest image-forming eyes evolved, not to detect objects, but to detect their motion.

A toad catches flying insects with its tongue; if by mistake it swallows a seed, a leaf, a piece of lint, it simply turns its stomach inside out and licks the offending particle off with its tongue.

Most animals live lives of unutterable regularity, and their eyes have evolved simply to warn them of interruptions to their usual programme. Rabbit eyes are remarkably attuned to motion, and this makes rabbits incredibly efficient at spotting predators. Otherwise, they are purblind: they live in a perpetual fog.

Humans are foragers; we take a more than usual interest in what things are. But even our eyes are tuned, first and foremost, to motion. In 1875 the Viennese physiologist Sigmund Exner showed that two brief, stationary flashes, provided they are not too far away from each other, are seen as a single object in motion. This habit of fusing stationary dots into moving objects makes a great deal of sense in nature, where prey and predators disappear and reappear constantly, as they move through grass, run behind trees, and peer around rocks. But the power of the phenomenon (called the phi phenomenon) will perhaps best be demonstrated the moment you set this book down and turn on the television. Every film and television programme ever made depends on phi. Both display still images quickly enough for our eyes to read them as a single moving image.

This is not 'persistence of vision', which is simply the eye's inability to tell a steady light from one that flickers faster than fifty times a second. In bright light, the eye does better; detecting flicker up to one hundred times a second. This is one of the reasons the front row of a cinema isn't popular: from here, bathed in light, and with much of the image located in the movement-sensitive periphery of vision, the eye is constantly aware of the flicker rate of the film.

When the flicker is obvious – not just snatching at the corners of your eyeballs, but there in front of you, a part of the cinema experience – it very quickly ceases to matter. Watch a silent film at its proper speed, on old equipment, and the image flickers about eighteen times a second. But, thanks to the phi phenomenon, the illusion of a single, moving image is still preserved, and the film is still eminently watchable. The phi phenomenon is quite happy to stitch together stills into moving images, even when they move quite slowly.

*   *   *

Take a moment to scan your surroundings: turn your head slowly from side to side. You'll notice that however smooth your head movements are, your eye 'snaps' from location to location. It's trying to keep whatever you are looking at steadily in view.

Why, when they compensate for our movements, do our eyes insist on *snapping* from position to position? Try moving your eyes across this sentence at a slow, even speed. Impossible, yes? The eye stutters. The only way you can remotely manage to move your eyes smoothly is by letting the page go out of focus – which rather defeats the point. If the eye's involuntary movements exist to compensate for our body movements, why can't the eye move as smoothly as the body does?

Actually, it can: slip a pen behind this book, so that the top pokes above the page. Move the pen smoothly back and forth. Follow it with your eyes. Suddenly, it's like someone has oiled the mechanism; the eyes move smoothly back and forth. This is the 'optokinetic reflex'. It is yet another visual stabiliser, one that Purkinje did not have the tools to unpick, since it uses visual information itself to move the eyes and stabilise the view upon the retina. It is the biological equivalent of the 'judder control' on digital video cameras.

Hold your eyes still, and the image of a moving target is quite likely to flash by individual photoreceptors in milliseconds. If it moves past very quickly, your photoreceptors won't have a chance to respond, and you won't see the target at all. Even if it's moving quite slowly, the target is likely to be blurred. This is why humans fixate so strongly on objects moving across their field of view – if they didn't, they wouldn't be able to see them. If you concentrate on a moving object, the eye will turn smoothly in an attempt to track it. The more you are interested in what you are looking at, the more it will try to hold on. But there is a limit: move your pen too fast, and the image will blur. The moment that happens, the eye will give up its 'tracking shot' and revert to 'angle shots'. It will *saccade*, snapping from location to location, capturing a selection of 'stills'.

In French, *le saccade* describes the way a sail snaps in the wind. (It was a Frenchman, the eminent vision researcher Louis Emile Javal, who named these sudden, involuntary movements of the eyeball.) There is nothing subtle about a saccade; nothing delicate. The movement is sudden, ballistic, and powerful enough that the eyeball commonly overshoots its target; a barrage of smaller correcting saccades follows any large eye movement.

What if our eyes did not saccade? What if they were entirely passive, so that we simply drank in the scene before us? What would we see? To get at the answer, it is not enough simply to stand and stare. Even when you are looking intently at something, your eyes are drifting and trembling three or four times a second. To still these tiny movements (called microsaccades) the eye has to be paralysed.

There are a couple of ways of disabling the eye. One method I absolutely would not recommend was dreamt up by the man generally regarded as Purkinje's intellectual successor, the physicist and philosopher Ernst Mach. As a student in Vienna in the 1850s Mach, unable to afford experiments in pure physics, caught the experimental psychology bug. His youthful enthusiasm for the fertile new zone he saw emerging between physics, psychology and physiology is evident from this experiment: he thought it would be a good idea to stuff putty under his eyelids to stop his eyes from saccading. The trouble with invasive methods like these – aside from the obvious dangers – is that the sheer discomfort and oddness of the procedure gets in the way of decent observation, as happened in Mach's case. A much more elegant approach became possible once plastics were developed and contact lenses became feasible. In the early 1950s, two research teams (one from Brown University on Rhode Island, and one from Reading University in England) stumbled independently on a way to stabilise images on the retina. They fastened tiny spotlights to the contact lenses their volunteers wore. It didn't matter how much the volunteers moved their eyes: the little spots of light would always be stabilised at exactly the same places on their retinas. The results could not have been more spectacular: after about a second of this curious, frozen vision, the volunteers lost sight of the lights.[22]

The eye exists to detect movement. Any image, perfectly stabilised on the retina, vanishes. Our eyes cannot see stationary objects, and must tremble constantly to bring them into view.

Three pairs of muscles move the eye, but nobody quite knows how. The superior rectus and inferior rectus (Labelled D and E in David Hosack's exquisite illustration (see colour section) move the eye up and down; the medial rectus and lateral rectus (C and A) move it from side to side. However, even after 450 years of anatomical research, beginning with Vesalius and Falloppio, there is still no general agreement on what the third pair of muscles is for. This pair, the superior oblique (B) and the inferior oblique (not shown), only came to notice some hundred years after the other two pairs had been described. In theory, the obliques should, via a system of pulleys, spin the eye about its axis – which is ridiculous enough. Making a model that can reveal their true function is far from easy. The earliest mechanical models treated the eye as a perfect sphere, and used strings for muscles. They did not reveal very much. For a start, all the muscles work in concert, so even when the whole problem is reduced to a matter of ball and string, the mechanics are horrendous. More seriously, muscles are not strings. They change their behaviour as they are stretched, and this means that they behave differently every time the eye looks in a different direction.

The puzzle of how the eye muscles pull the eye into position is more than a mere mechanical curiosity. Misalignment of the eyes is a treatable condition, and the more information we have about how the muscles are supposed to work, the better our corrective treatments become. With the advent of computers, it seemed as though the problem would be solved. David A Robinson, professor of biomedical engineering at Johns Hopkins University in Baltimore, attempted the first computer simulation, factoring in points of muscle origin and insertion, normal elasticity and muscle tone, the checks imposed by ligaments and other surrounding tissues, and much else. When he ran the program, the computer promptly turned its eyes to face the back of its head.[23] Robinson persevered. His model is good enough these days to be used all over the world to plan eye surgery. No sooner did the moving eye begin to make sense, however, than a new mystery arose: a human eye that did not move at all.

In 1995 John Findlay, from the University of Durham, and Iain Gilchrist, in Bristol – men who had built their careers on research into eye movement and vision – were disconcerted to hear from an undergraduate, AI, who claimed not to be able to move her eyes.[24]

AI had volunteered to take part in a study into people with squint – a persistent misalignment of the eyeballs. Her own squint was, she said, very mild, as it had been corrected by surgery when she was twelve. And there was one other thing she felt she should share with the researchers because it might disqualify her from the study: she had no eye muscles.

According to the surgeons who had operated to correct her squint, the muscles were so atrophied that they could never have worked. And this, for Gilchrist and Findlay, was no small source of worry. According to their theory of active vision – a model to explain why the eye has to move in order to see – AI's eyes should not have been able to function. And yet here she was: a student at university, writing letters, reading books, getting through life perfectly well and suffering from no apparent visual handicap. She did not drive, nor care for sport, but these life choices did not seem driven by any pressing or obvious visual defect. Though her eyes could not saccade, she could read normally. A speed of 257 words per minute wouldn't win her any speed-reading contests, but it was well within the range considered normal.

However, there was something about her: she moved like a bird. She twitched. She cocked and bobbed her head. The more closely her head movements were studied, the more bird-like they turned out to be. And for a very good reason: birds saccade mainly by moving their heads.

Not every eye in nature is moveable. Many – most – animals have eyes that are fixed permanently in place. Invertebrates often have eyes that cannot move, welded to a hard exoskeleton. A few get around this problem by locating their eyes on the end of long stalks. The jumping spider has come up with a different solution: it moves its retina, looking out through the lens of its eye the way a pilot looks through the glass canopy of his cockpit. For most insects, a fixed eye is not a problem: they are so tiny, they can simply swivel their whole bodies to change the view. Small animals are very powerful relative to their weight, so having to turn around is no real imposition. Airborne insects can simply swing about in space to capture any view they want.

The larger you get, the weaker you are, relative to your size. So larger animals move less of themselves to change the direction of their gaze. All vertebrates have eyes that are at least a little bit moveable. The smaller ones – birds, for example – make little use of this ability, and prefer to move their heads. Anything larger moves its eyes a great deal, and turns its head only when strictly necessary.

It appeared that AI, unable since infancy to move her eyes, had learned to saccade the way a bird does – by moving her head.

More strange has been the discovery that a bird's head movements – and even an insect's body movements – exactly parallel the movements of our own eyes. Birds and flies use the same basic strategies of smooth pursuit and saccade-and-fixate that we do. Indeed, these strategies are proving to be commonplace throughout the animal world.

Smooth pursuit makes sense when tracking a target, whether you're an insect, a bird, or a human being. Equally, by cutting up the visual continuum into a series of stills, the saccading eye (or head, or whole body) minimises motion blur, reduces disorientation, and more reliably reveals moving objects against a still background. The apparent eccentricities of human eye movement have turned out to be near-universal strategies for seeing.

4 – The 'problem' of consciousness

DB's head was a bomb, ready to explode. The malformed blood vessels in his brain could rupture at any moment, and destroy the delicate tissues surrounding them. There was no time to lose.

In 1986 a surgical operation removed the offending tissue from DB's brain. The price he paid for this life-saving surgery was high: by cutting into his right primary visual cortex, the surgeons inevitably damaged his vision. He could still see out of both eyes – but the left halves of both views were now invisible to him.

To appreciate how this could happen, we need to understand something about how we integrate the output of two eyes into a single view of the world. Take a playing card, tear it lengthways in two, and turn each piece upside-down. To fit the two halves together again, they have to be swapped around. Rays of light passing from the air into the dense, wet material of our eyes are refracted, or bent, so that the images that fall upon our retinas are upside-down. Like the two halves of the playing card, the images from the left and right eyes have to be swapped around before they can be brought together again at the primary visual cortex. This cross-over is accomplished before the optic nerves reach the brain proper, at a 'junction box', the optic chiasm.

Because human eyes face forwards, there is a considerable cross-over between the views of the left and right eyes, so it is not necessary for the optic nerves to swap sides entirely. Instead, nerve fibres carrying the left-most fields of view from both eyes are drawn together at the chiasm and continue to the right-hand side of the primary visual cortex. Nerve fibres carrying the right-most field of view of both eyes supply the left-hand side of the cortex. The result, in the primary visual cortex, is a single, inverted view of the world.

By damaging his right primary visual cortex, the surgeons were damaging DB's view of the left-hand side of space. That was what DB's doctor, Larry Weiskrantz, assumed. Then one day, DB's ophthalmologist, Mike Sanders, leaned into the left-hand side of DB's field of view during a test – and DB reached out to him.

Could DB see the hand? Certainly not. DB was a conscientious subject, and together he and Sanders had very precisely mapped the area of his blindness. But if DB couldn't see, who – or what – had guided his hand? Weiskrantz and Sanders set about testing the quality of DB's blindness. When they asked him whether a stick held in his blind area was horizontal or vertical, DB could not see the stick, and could only guess. The odd thing was, DB's correct guesses were more frequent than his mistakes. The difference between correct guesses and incorrect guesses was large – much larger than could be explained by chance. DB could not see the stick – but something within him could sense it.

What if they waved the stick around? Suddenly, DB's testimony changed. 'DB "sees" in response to a vigorously moving stimulus, but he does not see it as a coherent moving object, but instead reports complex patterns of "waves".'[25] As reported by Weiskrantz, a slowly-moving stick gives DB 'a kind of "feeling" that something is there.'

There's something a bit odd about the inverted commas Weiskrantz has placed around 'see'. It's hardly surprising that DB, half-blinded by his operation, was unable fully to appreciate a coherent image of a stick waving about on his blind side. What he did appreciate was a general perception of movement, rather as we might glimpse, 'out of the corner of our eye', a person or object passing through our peripheral vision. It would be nice to be able to say that some visual pathways are conscious, and some are unconscious; that consciousness was some kind of mysterious juice that flows along some of the brain's channels and not along others. But DB's experience shows us that things are not so simple. Weiskrantz's accounts show us a man who, given the right stimulus, becomes aware of being able to see.

Many eyes see in this limited fashion. Just because the tiny planktonic feeder *Pontella* has three lenses in each eye, this doesn't mean it sees the world in incredible detail. On the contrary, these three lenses have evolved to project light, very carefully, on to just six photoreceptors. If this is vision, my car alarm can see.

On the other hand, the study of some animal eyes has raised the unsettling possibility that these animals might be much cleverer than we thought they were.

Thanks to the work of the American psychobiologist Roger Sperry, in the 1960s, we are familiar enough with the idea that the left and right sides of our brain handle different tasks. Our 'right brains' are reputedly more visual and intuitive, whilst our left brains are more verbal and analytical. Until very recently, we thought this specialisation was unique; a sign of superior, human intelligence. We were wrong: in 2000, an Italian team, headed by Angelo Bisazza, found that mosquito fish (*Gambusia holbrooki*) prefer to check out likely mates with their left eye and possible predators with their right.[26] Evolving from an early visual specialisation, which helped our fishy ancestors recognise members of their own species, handedness appears now to be a common characteristic of all vertebrates.

However, when it comes to foxing our assumptions about intelligence, nothing comes close to the box jellyfish *Chironex fleckeri*. The venomous *Chironex* (its sting kills within four minutes) has eight camera-type eyes like our own, eight slit-shaped eyes and eight lensless pit eyes. That's two dozen eyes, connected, not to a brain, but to a simple nervous circuit. The eyes appear to be feeding the jellyfish information it hasn't the wit to use.

Is *Chironex*, with its fancy eyes, like the elk, whose impressive antlers are too big and heavy for its head? Is it caught up in some strange, highly specialised arms race that favours sophisticated eyes for their own sake? As we learn more about the behaviour of the box jellyfish (they are active, agile swimmers and the only jellyfish which copulate), an even more unlikely possibility gains credence. It seems that *Chironex* is thinking. Understanding how such brains might work – brains so unlike our own – is one of our great contemporary challenges.

TWO

The chemistry of vision

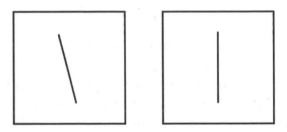

Natalie eats butter like it's going out of fashion. Bread, for her, is a butter delivery system. Salted butter will not do: there is too much water in it. What she wants is unsalted butter – the high-fat stuff that only the French manage to stay thin on. We give her slices of butter the way other people give their children slices of apple. She is an active child, and if we can encourage her to stay that way, she can have all the delicious, fatty, vitamin-rich butter she can cram down her throat.

Today, 254 million pre-school age children are deficient in vitamin A: just under half the children in Africa and just under three-quarters of the children of south-east Asia. At least a million child deaths each year are linked to vitamin A deficiency. Of the survivors, between a quarter and half a million go blind. The testimony of one field doctor, H A Oomen, speaking at a conference on vitamin deficiencies in 1958, is – fifty years on – still frighteningly topical:

> The over and over repeated experience of discovering a child, recently blinded, in the arms of the mother; having to tell her that I now could do nothing more to save its eyesight; remem-

bering that I could have done so with a few spoonfuls of cod-liver oil some days ago; these things still enter my nightmares.[1]

There is a chemistry to vision, and unpicking that chemistry changed the world. Most importantly for us, it opened up a new way of studying evolution. Most importantly for the health of children everywhere, it revealed vital facts about metabolism and nutrition. It inspired developments in forensic science.

It even got people killed.

1 – The tell-tale eye

'My boy, I wish you to witness an experiment.' He drew from its case a powerful microscope of French make.

'What on earth are you going to do, sir?'

The doctor's brilliant eyes flashed with a mystic light as he replied: 'Find the fiend who did this crime – and then we will hang him on a gallows so high that all men from the rivers to the ends of the earth shall see and feel and know the might of an unconquerable race of men.'

Thomas Dixon Jr, *The Clansman* (1905)

Odile Gaijean was a martyr to her son's fecklessness. During the 1980s Odile, a resident of eastern Alsace, ran up huge debts to bail her son out of a financially disastrous stay in Dakar, the capital of Senegal. When he turned up on her doorstep and announced that he was married, his mother was less than impressed. Then his wife – a former prostitute, Fatou Sarre – turned up.

Domestic bliss it wasn't. The feud between wife and mother-in-law began with strong words. Soon, they were trading blows. One day Odile, beside herself, picked up a knife. Fatou, not to be outdone, picked up a hammer.

Fatou won.

When it was done, she dragged her mother-in-law out of the house and into the barn. Afraid of what might happen if her husband found out it was she who had killed his mother, Fatou set about destroying the evidence of her crime. In March 1990, on trial for murder, Fatou explained to the court that she did not dare leave Odile's dying vision intact, for it would surely show her brandishing the hammer.

Which was why she gouged out her mother-in-law's eyes.

Three years later, there was a similar case, involving a hapless Cameroonian housebreaker. The press detected a pattern. Was France witnessing a reawakening of ancient 'Cameroonian legends' and 'African beliefs'?

No. On further interrogation, it turned out that Fatou's 'beliefs' – such as they were – dated back to a Bollywood film she had seen one night in Dakar. It was a whodunnit, loosely based, not on 'ancient legends' (Cameroonian or otherwise) but on a much more recent source – European detective fiction.[2]

The short story, *Claire Lenoir*, caused a minor sensation on its publication in Paris in 1867. Huysmans refers to it in his novel *A Rebours*, and twenty years later, there was still enough mileage in the idea for the author, Villiers de l'Isle-Adam, to expand it into a successful novel.

Claire Lenoir is the wife of Césaire, an old friend of the story's narrator, Tribulat Bonhomet. On his way to see them, Bonhomet befriends Henry Clifton, a young English naval lieutenant who has taken a posting in the South Seas to cure himself of an ill-fated love affair with a married woman (who bears more than a passing resemblance to Claire).

Later – just before he meets the Lenoirs – Bonhomet stumbles over a news report:

L'Académie des Sciences de Paris has stated the authenticity of certain surprising facts. It can be asserted that the animals destined to our nourishment – such as sheep, lambs, horses and cats – conserve in their eyes, after the butcher's death stroke, the impression of the objects they have seen before they die. It is a photograph of pavements, stalls, gutters, of vague figures, among which one almost always distinguishes that of the man who has slaughtered them . . .[3]

It is an ill omen and, sure enough, things rapidly take a turn for the worse. His friend Césaire falls desperately ill. Somehow Césaire has discovered his wife has been unfaithful; on his death-bed, he swears a terrible oath of revenge.

A year later, news arrives that Henry Clifton has been brutally murdered by a vampiric assailant. Césaire's widow Claire takes to her bed, driven out of her mind by persistent nightmares. Bonhomet is with her when she dies and, noticing a blurry image lingering in her eyes, he examines them with the aid of his trusty ophthalmoscope. What he sees:

> no language, dead or alive . . . could, under the sun and under the moon, express in its unimaginable horror . . . I saw the skies, the far-off floods, a great rock, the night and the stars! And upright, on the rock, larger in height than the living, a man . . . lifted with one hand, towards the abyss, a bloody head, with dripping hair . . . and, in the severed head, the features, frightfully obscured, of the young man I had known, Sir Henry Clifton, the lost lieutenant!

It is strange, at this remove, to imagine the popular curiosity generated by the ophthalmoscope; a device, combining magnifying lenses and a light-source, with which one can view the inside of the living eyeball. The nearest contemporary equivalent would, I suppose, be the craze, a few years ago, for the 'magic eye' posters that conceal three-dimensional figures in a computer-generated pattern of apparently random blobs. They too, as we have seen, have their serious scientific side, and now that the fashion for them is past, like the ophthalmoscope, they have largely slipped out of the public gaze, back into the somewhat rarefied world of vision science.

The way de l'Isle-Adam collides the two great scientific crazes of his day – ophthalmology and photography – is a splendid example of the fantastic at work. The eye is the soul's camera, and the ophthalmoscope reaches in to spy upon the soul!

If the idea seems trite today, it has only become so through repetition. For de l'Isle-Adam's readers, *Claire Lenoir* addressed the bodily anxieties generated by the era's huge advances in surgery and imaging, as surely and as disconcertingly as the the films of David Cronenberg address the bodily anxieties of our own, increasingly anonymous and 'virtualised' day.

Exposed, the body reveals its affinity with mechanism: the eyes are cameras, and to the well-equipped observer they reveal the secrets of the heart as surely as a daguerrotype reveals the interior of the sitter's drawing room. This is a great opportunity, but a risky one, at least in the uncanny world of *Claire Lenoir*. At some level, humans are not machines; they are messy, unpredictable, generators of nightmares. You can look, but you may not like what you find:

> And Science, the old queen-sovereign with clear eyes, with perhaps too disinterested a logic, with her infamous embrace, sneered in my ear that she was not, she also, more than a lure of the Unknown that spies on us and waits for us – inexorable, implacable!

Such was the public's fascination for the new-fangled excitements of forensic science, that the fiction which followed *Claire Lenoir* become steadily less like ghost stories, and more like police procedurals. In 1897, Jules Claretie wrote *L'Accusateur*, in which the eye of a murdered diplomat, dissected by the police, reveals the blurred image of the face the victim saw as he died.[4] Jules Verne's *Les Frères Kip* (1902) sits even more comfortably within the detective genre in its tale of innocent brothers, arrested for a murder they did not commit, released at the eleventh hour on the evidence of a retinal photograph. Three years later, retinal photography revealed the identity of a rapist in Thomas Dixon Jr's jaw-droppingly racist novel *The Clansman* – the basis for D W Griffith's film *The Birth of a Nation*.

Philosophers and scientists could no more resist the analogy between vision and photography than could the reading public. In 1859, Charles Darwin's *The Origin of Species* explained how, given large numbers, time, and a testing environment, living things acquire an efficient, apparently designed form. In acquiring vision, had nature anticipated the very solutions hit upon by those pioneers of photography, Wedgwood, Niépce and Daguerre?

For the retina to behave like a camera's plate, it had to contain some light-sensitive chemical, analogous to the silver nitrate film covering the glass slides on which the earliest photographs – daguerreotypes – were captured. The presence of a reddish pigment in the retinas of frogs and squid was established in 1851 by the German anatomist

Wilhelm Kühne.

Heinrich Müller (1820–1864). In 1876, his brilliant young compatriot, Franz Christian Boll (1849–1879), narrowed the pigment's source down to one particular group of cells – the 'rods'. The unusual shape of these cells, coupled with the way they were packed so closely together across the retina, was a clear indication that they possessed an optical function. Were these the cells that saw?

Boll noticed that, when a frog died, the reddish colour of its retinas bleached away, so that after about a minute, they appeared colourless. At first he thought this was a consequence of the frog's death. But then he discovered that even dead frogs retained healthy reddish retinas for up to twenty-four hours, provided they were kept in the dark. If light bleached a pigment stored in the 'rod' cells of the frog retina, might that chemical event not convey the impression of light to the frog's brain?

In 1879, Boll's experiments were cut short: he was only thirty when he died, a victim of the tuberculosis that had blighted his otherwise glittering professional life. But he had done enough to convince his peers at the Berlin Academy that 'This change in the outer segments of the rods forms indisputably a part of the process of vision.'⁵

Among Boll's admirers was Wilhelm 'Willy' Kühne (1837–1900). A powerful figure in the field of physiology, Kühne had recently succeeded Hermann von Helmholtz as professor of Physiology at Heidelberg. Kühne took up Boll's discoveries with 'fiery zeal'.

Boll had called his pigment 'visual red'; Kühne reckoned it was more purple, and changed the name accordingly. Otherwise, Kühne agreed with Boll – and acknowledged his debt – on almost every important point. If the retina was the eye's 'photographic plate', then Boll was surely right, and 'visual purple' was the eye's equivalent of the camera's silver nitrate. Kühne hoped to complete Boll's work by extracting evidence from the retina itself.

Working with the simplest equipment in a darkroom, Kühne found a way to maintain retinas so that he could study them closely. In the dark, the dissected retinas remained purple. Light bleached them, but not instantly. Kühne observed distinct stages in the bleaching process, from purple to orange, yellow, and buff, until the retina became entirely transparent.

Fifty years passed before any new knowledge was added to the meticulous description of visual purple assembled by Kühne over two busy years. Indeed, the pile of unanswerable questions Kühne had amassed concerning the stuff might have obscured his actual achievement, had he not gone to extraordinary and repeated lengths to publicise his central finding: that the eye behaves like a camera.

On November 16, 1880, in the nearby town of Bruchsal, a young man was beheaded by guillotine. The body was taken to Heidelberg, where Kühne, in his capacity as Professor of Physiology, was waiting. In a gloomy room, its few windows screened with red and yellow glass, Kühne dissected the dead boy's eyes. Ten minutes later, he showed his colleagues a sharp pattern on the surface of the left retina. This, Kühne told the company, was an optogram: a dying vision, preserved as a pattern in that delicate and unstable pigment, visual purple. The trouble was, Kühne's sketch failed to match any object that may have been visible to the felon as he died.

We know now that Kühne had been confronted – and bested – by a curious specialism of the human eye. Its area of focused vision

A rabbit's dying vision, captured by Kühne's most successful optogram.

– the fovea – is tiny in comparison to the size of the whole retina. Even more puzzlingly, the fovea has very few rods. Kühne had better success recording the optograms of frogs and rabbits. This account of Kühne's early and modestly successful 'optography' is by the biochemist George Wald – himself a Nobel prize-winner for his studies of visual pigments. (Wald refers to visual purple by its currently preferred name, rhodopsin.)

> One of Kühne's early optograms was made as follows. An albino rabbit was fastened with its head facing a barred window. From this position the rabbit could see only a gray and clouded sky. The animal's head was covered for several minutes with a cloth to adapt its eyes to the dark, that is to let rhodopsin accumulate in its rods. Then the animal was exposed for three minutes to the light. It was immediately decapitated, the eye removed and cut open along the equator, and the rear half of the eyeball containing the retina laid in a solution of alum for fixation. The next day Kühne saw, printed upon the retina in bleached and unaltered rhodopsin, a picture of the window with the clear pattern of its bars.[6]

The key word in this passage, of course, is *fixation*. A living retina has no interest in fixing images. Far from capturing perfect stills, it

offers up a constantly changing view. As Kühne himself wrote; 'the
retina behaves not merely like a photographic plate, but like an entire
photographic workshop, in which the workman continually renews
the plate by laying on new light-sensitive material, while simultaneously erasing the old image.'

It is impossible now to say whether Kühne's macabre autopsy on
the felon from Bruchsal was a rational and necessary extension of
his laboratory work, or an aberrant piece of showmanship. His work,
and the work of his predecessor Franz Boll, had been widely
published in magazines across the globe. In Britain, *Fortnightly
Nature, The Athenaeum* and *The Nineteenth Century* ran articles, as
did *Harper's Weekly* in America. And we should not under-estimate
the pressure Kühne and other researchers were under to extend the
frontiers of forensic science. Since 1857, when the first, probably
apocryphal story of optography appeared in *Notes and Queries*,
curiosity about the possibilities of forensic photography regularly
found its way into court transcripts. By 1869, the French Society of
Forensic Medicine was so concerned that the courts might start
admitting 'optographs' as evidence in murder trials, it asked Dr
Maxime Vernois to conduct a scientific study. Vernois set about his
task with enthusiasm, 'violently' slaughtering seventeen experimental
animals 'under bright lights'. They died in vain:

> It is impossible to find upon the retina of a victim the portrait
> of its murderer or the representation of whatever object or physical
> trait that presented itself to its eyes at the time of death.[7]

Despite the failures recorded by Vernois and Kühne, enthusiasm for
the idea persisted, and spiced up genuine murder investigations. In
1888 Walter Dew – a London policeman later famous for catching
the murderer Dr. Hawley Harvey Crippen – witnessed 'the most
gruesome memory of the whole of my Police career' when he entered
13 Millers Court in Whitechapel, where lay the disordered remains
of Mary Kelly, a victim of Jack the Ripper. 'Several photographs of
the eyes were taken by expert photographers with the latest type
cameras,' Dew remembered in his autobiography, in the 'forlorn hope'
that an image of her killer was retained on the retina.[8]

Is Dew's memory reliable? Perhaps not: but the idea was certainly on everyone's mind. At the inquest of Annie Chapman, another of Jack's victims, the jury foreman asked police surgeon Dr George Bagster Phillips, 'Was any photograph of the eyes of the deceased taken, in case they should retain any impression of the murderer?' 'I have no particular opinion upon that point myself,' Phillips replied. 'I was asked about it very early in the enquiry and gave my opinion that the operation would be useless, especially in this case.'

There is, as far as I have been able to ascertain, only one, extremely dubious, example of a conviction attained by an optogram. It is from the *Sunday Express* and is quoted in Veronique Campion Vincent's invaluable paper 'The Tell-Tale Eye', published in the journal *Folklore* in 1999. It seems that readers of the edition published on 4 July 1925 thrilled to the tale of Fritz Angerstein, a resident of Limberg in Germany, who killed eight members of his household with a hatchet. The police, seeking a quick confession, told the court that they had taken a photograph of the retinas of Angerstein's gardener, which revealed 'a picture of Angerstein with his raised arm gripping a hatchet.' Hearing this, Angerstein threw in the towel and confessed to the killings.

And the police photograph? Most likely, it was simply a tale, spun to demoralise a gullible defendant. It was never made public.

2 – A question of diet

Nutrition has often been the subject of conjectures and ingenious hypotheses but our actual knowledge is so insufficient that their only use is to try to satisfy our imagination. If we could arrive at some more exact facts they could well have applications in medicine.

                                        Francois Magendie[9]

Robert Boyle (1627–91), the fourteenth child of the fabulously wealthy Earl of Cork, never needed – and never bothered to acquire – gainful employment of any kind. Instead, he became one of seventeenth century Europe's leading intellectuals. Boyle was a key advocate of experiment as a path to truth. 'Our Boyle,' his friend Henry Oldenburg wrote, with wry wit, 'is one of those who are distrustful enough of their reasoning to wish that the phenomena should agree with it.' But there was another side to Boyle, which could not resist a good tale or a wild claim, provided it was well-expressed and intelligently defended. These tall stories pepper his writings and correspondence, testifying not only to the sharpness of his intellect but also to the breadth of his imagination.

Consider this interview with Major Edward Halsall. (Boyle grants Halsall anonymity in his account, but we can be fairly sure it was him, thanks to the researches of the historian Michael Hunter, who brought this story to my attention.) Halsall was a royalist who was imprisoned during the Cromwell era for his part in the murder of Anthony Ascham, the Commonwealth's envoy to Spain, in 1650.

Halsall, 'being accused at Madrid of a state crime, had irons put upon his feet, and was kept for twenty months in a room, which though otherwise not unfurnished was, by reason of its darkness, reckoned among the dungeons, for there was no window to it.' During that time 'he saw no light at all, save when the gaolers came from time to time with candles to bring him wine and other provisions, and see that he made no attempt to escape'. It took Halsall's eyes seven months fully to adjust to the dark, and even then he saw things only dimly. By the end of his imprisonment, however, Halsall's night vision had grown prodigious: '. . . he could see the mice that used to feed upon his leavings plainly run up and down the room' and 'he could see well enough to make a mousetrap with his cup.'[10]

Halsall claimed that, had he been given books and paper, he would have been able to see well enough to read and write. Unlikely as this is – the discrimination of print requires considerably more light than the discrimination of objects – everything else about his account seems eminently plausible.

It is extremely unlikely that anyone with access to this book – including me – has the remotest experience of the real capabilities of night vision. We live our lives doused in light, and we think of our night vision as somewhat second-rate, compared to the night vision of cats or foxes or owls. But although we lack certain exotic adaptations that can aid nocturnal vision, we can still detect light that is a billionth the strength of daylight. This is sensitive enough that, in good conditions, we could see the flame of a single candle seventeen miles away. Very few species could do better. (It is our relatively poor hearing and sense of smell that lets us down: nocturnal species rely very much on these senses to supplement their vision.)

Like so many of our remarkable physical and sensory abilities, night vision is really appreciated only when it goes wrong. Night blindness is perhaps the most venerable of all eye diseases; it has excited medical opinion for at least as long as cataract.

We can be reasonably certain that 'night blindness' has been known in Egypt, India and China since records began. In all three cultures, the cure seems to have been roasted or fried animal liver. This excites today's historians, since we know liver is a good source of vitamin A, and vitamin A plays a central role in vision.

The Berkeley nutritionist and historian George Wolf is prepared to stake a claim for the ancient Egyptians, who – according to the Ebers papyrus of 1520 BC – treated an inability to see at night by squeezing the juices from an animal's liver straight into the patient's eyes.[11] Since the liver stores about ninety per cent of the body's vitamin A supply, this gruesome-sounding procedure would work. Any liquid of that sort passing down the naso-lacrimal duct would be absorbed into the blood stream, as the makers of eye-drops well know. To this day, both night blindness and painfully dry eyes (the two conditions go hand in hand) are treated with eye-baths of fish liver oil.

Wolf's case is not altogether sound, because the Ebers papyrus – the earliest medical text we know of – lists illnesses only by name. Is 'night blindness' a lack of night vision, or the inability to see *except* at night (photophobia)? Or is it neither? There is no way to be sure. Wolf's detractors go further, pointing out that in early medical texts, fried or roasted animal liver seems to be the cure for everything. They also point out the risk of reading magical texts through medical spectacles. The appearance of the liver is affected by numerous illnesses and conditions, and this variable appearance has made it especially useful to soothsayers throughout the ages, who descried, in that lump of protean flesh, everything from the outcome of an eye infection to the propitious date for a war.

If the Egyptians did know how to cure night blindness, the knowledge was lost: Aulus Cornelius Celsus, the Roman encyclopedist, born 25 BC (about 1,500 years after the Ebers papyrus was written), makes no mention of liver or fish oil in his otherwise spot-on description of chronically dry eyes. Also, he makes no mention of night-blindness:

> There is a kind of dry inflammation of the eyes called by the Greeks xerophthalmia. The eyes neither swell nor run, but are nonetheless red and heavy and painful and at night the lids get stuck together by very troublesome rheum . . .[12]

It all sounds innocuous enough. Celsus – an encyclopedist, rather than a medical man – saved his readers from the more unpalatable details of the diseases he described. To get a full measure of the seriousness of xerophthalmia, we do better to turn to a source closer to home. This is Colonel Robert Henry Elliot, (1864–1936), writing of his experiences as an English ophthalmologist in India:

> Never, whilst memory lasts, can one obliterate the mental pictures of those pitiful little bundles of marasmic, apathetic humanity, lying in the arms of gaunt women, on whose faces and forms famine had laid its devastating hands. Their faint, feeble, fretful wails ring still in one's ears today, summoning up visions of wasted stick-like limbs, of distended abdomens, of dry, inelastic, scurfy, scaly skins, of hair scanty, brittle and dry, and of sightless desiccated eyes . . .[13]

From Celsus's dry eyes to the full-blown deterioration of the cornea and, ultimately, the ghastly systemic failures described so movingly by Colonel Elliot, the distance is great. It is hard to believe that they are, in essence, degrees of the same condition. Even more bizarre is the notion that this condition has nothing to do with disease, nor with starvation. What other explanation can there be for the failure, first of vision, and then of the whole body?

In 1816, in an attempt to understand the biological value of certain foodstuffs, Francois Magendie – a child of revolutionary Paris and a former physician – set out to observe the effects of a restricted diet. He was particularly interested in what role nitrogen has to play in digestion. The answer he got back, after ten years of painstaking work, was none at all: 'As so often in research,' Magendie wrote, with what bitterness we may imagine, 'unexpected results had contradicted every reasonable expectation.' But in the pursuit of this knowledge, Magendie had stumbled upon a striking, if unpleasant, discovery: he had found that he was able to starve his experimental dogs to death on diets that should, on the face of it, have given them all the energy they needed for life.

By his own account, Magendie 'placed a small dog about three years old upon a diet exclusively of pure refined sugar with distilled water for drink; he had both *ad libitum*.' By the third week the animal, already weakened, lost its appetite, and developed small ulcers in the centre of each cornea. The ulcers spread, then the corneas liquefied. Shortly afterwards, the dog died.

Magendie tried other nutritious foods. 'Everyone knows that dogs can live very well on bread alone,' he confidently asserted – but when he put this to the test, he found that 'a dog does not live above fifty days.' The most calorific foods in Magendie's pantry – wheat gluten, starches, sugar, olive oil – were not enough for life. This was totally unexpected. There was something missing – something available only as part of a varied diet – but what?

It was the Japanese doctor Masamichi Mori (1860–1932) who drew serious medical attention to the problem. In his 1904 paper – the first account of a xerophthalmia epidemic and the first really good large-scale study ever made of dietary deficiency – Mori details the plight of 1,400 children between the ages of two and five. Some had night-blindness; others suffered dry and ulcerated eyes. Some had eyes that had completely disintegrated. Some were dying. They were suffering from xerophthalmia, known to Mori as 'hikan'.

Mori's study of hikan is an exemplary piece of medical detective work. Local lore said that inappropriately early weaning could bring on the condition; and sure enough, rural women tended to withdraw the breast much earlier than women who lived in towns, where hikan was relatively rare. At the same time, Mori noticed that the children of fishermen seemed strangely immune to the condition, even though they were weaned earlier than urban children.

Mori concluded that animal oils – and fish oils, especially – were essential to preserve health. The more fats you ate, the less likely you were to suffer from hikan.

As the nineteenth century waned and turned we find many insightful descriptions of the condition; among Russian peasants during severe Lenten fasts, children in a Paris orphanage, Russian soldiers in the Crimea, waifs and strays in Berlin, and even among the soldiers of the Confederate army in the United States, where the condition was blamed on the 'meagre diet, absence of vegetables and vegetable oils, and other depressing influences of a soldier's life'.

The trouble was knowing what to do about it. The great public health advances ushered in by Pasteur's germ theory had drawn attention away from questions of diet. The experience of the German doctor S Kuschbert is typical: in 1884, he noticed night-blindness and dry eyes in twenty-five of the orphan children with whom he was working. On further investigation, he discovered that the ulcers on the children's corneas contained bacteria. He drew the obvious wrong conclusion: that the children had caught one of Pasteur's famous germs.

Even allowing that some conditions were caused by a bad diet – rickets and scurvy being the best known – no one had isolated the factors that protected the body against them. These dietary factors had to be tiny – so tiny as to be ephemeral – and how could a body be so dependent on mere traces of something?

Incredulity ran so high that it took the evidence of a world war to overcome it. The maritime blockade of continental Europe during World War I led to malnutrition in several countries. Analysed by physicians and nutritionists, these cases held the clue to the greatest medical advance since Pasteur: the discovery of vitamins.

In 1917, Carl E Bloch (1872–1952) was the physician in charge of the paediatric clinic of the University of Copenhagen and of a children's home which cared for eighty-six infants under the age of two. The home had two buildings. The first cared for newborns and the sick. The other had two rooms, each housing healthy babies. As you would expect, the healthy babies enjoyed a more varied diet than the sick ones.

The outbreak, when it came, was sudden and devastating. Bloch reported forty cases of xerophthalmia among the children under his care. The children were sensitive to light, and in the early stages some were night-blind. Dryness of the conjunctiva followed, spreading to the cornea. He also reported 'a viscous discharge from the eyes'. All the cases occurred between May and June 1917, *among the healthy children.*

How could children on a varied diet fall victim to a dietary illness to which already sick children, fed on a narrow diet of gruel and whole milk, were immune? Bloch found the answer in the matron's meticulous records: following a minor attack of diarrhoea among the healthy children, she had swapped the traditional beer-and-bread soup, made with whole milk, for an oatmeal gruel with rusks.

And there it was: whole milk. Children deprived of whole milk succumbed to xerophthalmia. The others, even the ill ones, were immune.

Bloch wasted no time. Along with whole milk, he fed the xerophthalmic children other fats and oils. Cod-liver oil turned out to be especially effective: even apparently hopeless cases recovered within a week. The xerophthalmia never returned. A year later, when the German submarine blockade drove the prices of butter and whole milk well beyond the reach of the poor, the Danish government introduced butter rationing; every adult and child could obtain 250 grammes of the stuff every week. Xerophthalmia, which was never rare in Demark and which had reached epidemic proportions during the war years, virtually vanished overnight. Only when hostilities ended and butter was no longer rationed did it make a brief and short-lived resurgence.

The cluster of related compounds which travel under the banner 'vitamin A' were finally isolated between 1912 and 1914 by two independent American teams. In the twenty years that followed, thirty clinical trials of vitamin A established its importance, not just for eyesight, but for warding off a host of infections. Vitamin A, the 'anti-infective vitamin', was vigorously promoted by doctors and pharmaceutical companies. Not until 1947 was a way found to synthesise it; the inter-war generation had to resign themselves to daily doses of cod-liver oil, whose ghastly taste was often disguised with treacle.

Many scientists acquire their sense of vocation early. George Wald (1906–1997) was not one of them. Born in New York, the youngest son of Polish immigrants, Wald had always been good with his hands – he and his friends built a simple radio so they could listen to ball games – but a visit to an engineering firm where a friend of his father worked convinced him that engineering was not for him. After taking a vaudeville act around local Jewish community centers, Wald decided that his showmanship marked him out as a born lawyer. As

George Wald.

a pre-law student at New York University, he soon learned otherwise.

Eventually, he fell into zoology – and never looked back.

By 1932, the twenty-six year old postgraduate was on the first leg of an extra-ordinary tour of German universities, pursuing the mysterious relation between vitamin A and vision.

The adventure began auspiciously. Arriving in Berlin with his wife Frances, Wald went to work with Otto Warburg – a Nobel laureate, no less. Wald and Warburg knew that vitamin A played some role in vision because a lack of vitamin A brought on night blindness. But how direct was that link? Wald's first job was to confirm the results of a dramatic experiment performed by the Yale ophthalmologist Arthur Meyer Yudkin, who had fed the freeze-dried retinas of slaughtered farm animals to vitamin A-deficient rats. These rats recovered as quickly as rats given cod-liver oil – a long-established source of vitamin A. The conclusion – that retinas contain vitamin A – was confirmed by Wald's chemical tests.

Next, Wald went to Zurich, to the laboratory of the Russian-born chemist Paul Karrer, to find out whether vitamin A was a component of the retinal pigment, 'visual purple' or rhodopsin, that Boll and Kühne had already identified. Wald and Karrer collected retinas from cattle, sheep, and pigs, extracted them with solvents, and found vitamin A in all of them.

The final leg of Wald's journey – to the Kaiser Wilhelm Institute for Medical Research in Heidelberg – was meant to be the crowning event of his tour. There, he was to meet Otto Meyerhof, an expert in the biochemistry of muscle tissue. But Adolf Hitler had come into power only months before, in January 1933, and the Nazi authorities had begun to remove some of the finest minds in the country from their academic posts. In Heidelberg, the Institute was putting up a fine show of resistance; its president and director shared an unerring knack for losing official letters. Nevertheless, the local Nazi party was agitating for the removal of Meyerhof, who was, like Wald, a Jew.

Even as Wald was pushing paper around Meyerhof's laboratory, waiting for him and his team to return from holiday, Wald's backers, the National Research Council, made jittery by the deteriorating political situation, told him to cut short his trip and head home. At the same time, a delivery arrived. Someone must have got the dates wrong – the building was virtually deserted and here, stacked in the hall, were crates and crates of frogs. Three hundred frogs.

It was an opportunity George Wald seized with both hands. If he was not to work with Meyerhof, he could take advantage of the great man's extra-ordinarily well-equipped laboratory. He set about recreating the findings of Kühne, who had seen the purplish rhodopsin of frog retinas pass through distinct colour shifts on its way from purple to clear. First, light bleached 'visual purple' to something Kühne had dubbed 'visual yellow'. On further exposure, visual yellow then became transparent: Kühne called this stage 'visual white'. Wald performed chemical tests to find out what 'visual white' was made of.

It was pure vitamin A.

What of 'visual yellow'? Visual yellow was very similar to vitamin A, Wald found, but with one telling difference: in the dark, it converted readily back into rhodopsin. He had begun to uncover the biochemistry of vision.

Wald's account of the visual cycle goes something like this: the pigment rhodopsin consists of a membrane protein (an 'opsin') connected to a light-reactive component, 'retinal'. When light hits the rhodopsin pigment, the retinal changes shape and the pigment changes colour, from purple to yellow. In the dark, the retinal flips back to its former shape, and the pigment turns purple again. Over time, the retinal decays. Extra vitamin A is needed to support the re-combination of retinal and opsin into rhodopsin, and if the body is deficient in vitamin A, night-blindness follows.

Safely back at Harvard, Wald contemplated the life's work now stretching ahead of him. The questions his work threw up fell under

two broad headings. First, how prevalent was rhodopsin? The eyes
of frogs, sheep, cows, pigs and people all contained rhodopsin. Was
this a concidence, or was rhodopsin truly universal? Second, how
important was rhodopsin to vision? Maybe rhodopsin was just one
of many different pigments used by the eye. Maybe pigments were
just the easiest to spot of several different visual mechanisms. The
number of possible complexities was enormous.

It took Wald, his wife and co-worker Frances, his long-time collaborator Paul Brown and untold others who passed through their laboratory about twenty years to work through all the possibilities. In
that time, they found virtually nothing to complicate the original
picture: every single eye, with no exceptions, uses rhodopsin to see.
Some animals, it is true, do supplement rhodopsin with additional
pigments. None the less, every animal makes use of rhodopsin.
Rhodopsin is a universal visual pigment.

Complications did set in concerning rhodopsin itself. Wald and
his co-workers found that it comes in many different flavours.
Miniscule genetic differences in the opsin membrane can transform
the behaviour of the whole molecule, making it react most strongly
to quite different wavelengths of light. This is what powers colour
vision in the great majority of animals.

So it was that George Wald and his collaborators – in particular
his second wife, Ruth Hubbard – led the post-war revolution that
changed biology from a cellular to a molecular science. In his Nobel
lecture of 1967 he explained:

That early *Wanderjahr* in the laboratories of three Nobel laure-
ates – Warburg, Karrer, Meyerhof – opened a new life for me:
the life with molecules. From then on it has been a constant
going back and forth between organisms and their molecules
– extracting the molecules from the organisms, to find what
they are and how they behave, returning to the organisms to
find in their responses and behavior the greatly amplified
expression of those molecules.[14]

Molecules, like animals, have a story to tell. Molecules, too, have a
history.

THREE

How are eyes possible?

1 – Building blocks

Woods Hole, a town on Cape Cod, Massachusetts, has been a Mecca for America's marine scientists for over a century. Here, in the mid-1930s, at the town's Marine Biological Laboratory, George Wald pioneered a new way of looking at the story of life on earth: rather than trace the history of whole animals, why not trace the history of their ingredients? Why not trace the history of their molecules?

Wald's work at Woods Hole threw up as tidy an example of such a history as one could wish for. He discovered that amphibians and salt-water fishes both use a common form of rhodopsin – the same form used by the rods in our own retinas. Fresh-water fishes, on the contrary, use a form of rhodopsin capable of detecting longer wavelengths of light. (Water absorbs long-wavelength light, which is why, the deeper you swim, the more blue everything becomes. The light dappling the shallows where most freshwater fish swim has a longer average wavelength than light that has perhaps had to penetrate half a kilometre of ocean.) Salmon, which spawn in fresh water, use fresh-water rhodopsin; other species, which spawn in salt water, use salt-water rhodopsin. Tadpoles, Wald found, swap salt-water rhodopsin for fresh-water rhodopsin during their metamorphosis into frogs.

Wald concluded that salt-water rhodopsin is older than fresh-water, and that fresh-water rhodopsin has evolved from the salt-water form. When he wrote up these and similar findings in provocative articles with titles like *The Origin of Optical Activity*, they opened the door to a whole new field of evolutionary biology.

In 1990, an ambitious project was launched to compare the sequence of letters that spell out the genes for all the different kinds of rhodopsin then known. Genetic sequences had by then been established for four fruit-fly opsins, one opsin from an octopus, all four opsins from humans (one from the rod cells, and three from our three different types of cone) and one rod opsin each from chickens, sheep, cows, and mice. The results showed that these different opsins all descended from a common ancestor. In the 600 million years it had taken for animals as diverse as fruit flies and chickens and people to arise from that shared ancestor, rhodopsin – a molecule common to them all – had undergone only the most trivial changes.

These findings are impressive, but not unexpected. In 1882, the German physiologist Theodor Engelmann speculated that, for it to occur in the eyes of virtually every animal, rhodopsin must have existed at a very early stage of evolution, most probably before eyes themselves had arisen. Englemann's conjecture arose out of his work with photosynthesis: he had discovered that some bacteria are able to make food from water and the carbon dioxide in the air, using the energy from sunlight. Nearly a century later, in the late 1970s, Engelmann's hunch paid off: a form of rhodopsin, harnessed variously as a defence against ultraviolet light, and as a mechanism of photo-synthesis, was found in bacteria.

Rhodopsin's use in vision, which concerns us here, is a story that begins surprisingly early in the evolution of life. The ability to detect not just the presence of light, but also its intensity, wavelength and the direction it is coming from, is present among humble bacteria, fungi, algae, and even the single-celled protozoa.

One of these, *Euglena gracilis*, an algal flagellate, uses its photoreceptive pigments to acquire the energy to drive its whip-like tail. This 'propeller' is constructed by the complex folding of the cell membrane, and is arranged to drive *Euglena* towards the light. When *Euglena* enters sufficiently bright conditions, it begins to photosynthesise. When the light declines, the mechanisms of photosynthesis break down and *Euglena* survives by eating, making it truly both a plant and an animal. Though *Euglena* contains rhodopsin, its tail is driven by a different set of pigments, the flavins. Flavins are as widespread in nature as rhodopsin and have an analogous evolutionary history. They crop up in fungi, algae and plants and appear to have independently arisen at least three times. In the higher animals, ourselves included, flavins control the metabolism of the retina; they, too, have found a role in vision.

Euglena's pigment globules have no particular arrangement, and though they power a directional tail, they themselves resemble nothing more sophisticated than an 'umbrella' or light filter. The multiple pigment layers inside the alga *Chlamydomonas*, on the other hand, suggest a complex optical structure. The layers may work as reflectors

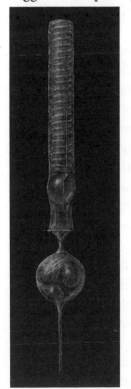

to contain and concentrate the light, increasing the likelihood that it will react with the pigment. Is this the world's most primitive retina? The protozoan *Erythropsis* boasts an even more sophisticated eyespot, topping its highly-ordered crystalline structure with a large, transparent, spherical lens.

The closer we come to vision proper, the more sophisticated the behaviour of an animal – even a single-celled animal – can be. In 1975, the Italian researchers Ester Piccinni and Pietro Omodeo were astounded to discover a group of algae which used their simple eyes to hunt for food. Lacking anything approaching a nervous system, these tiny

Gordon Lynn Walls's idealised rod cell; its main body and pigment stack are connected by a tail-like filament.

creatures could spot prey and entangle it with explosively released threads.

Evolution cannot un-invent. The basic building blocks of complex structures like the eye were available long before they were recruited for sophisticated tasks like vision. Human lenses are derived from proteins that protect bacteria from rapid rises in temperature. The flexible guanine mirrors which make a cat's eye glow in the dark also gas-proof the swim bladders of fish. The red and yellow pigments in the eyes of insects also colour the wings of butterflies. We have noted the slinky-like behaviour shared by retinal rods and muscle cells; it may be worth adding that the filament connecting the two halves of the retinal rod cell is built exactly the same way as *Euglena*'s tail.

Eyes have been around for at least 538 million years – but their constituent elements are more ancient still.

2 – The third eye

Light hits a rod in my eye, and bleaches the pigment within. A cascade of chemical events shuts down the rod's 'ion channels' – proteins that pass an electrical current around the cell. With its ion channels shut down, production of a chemical, glutamate, is stopped. This sudden chemical 'silence' excites a neighbouring nerve cell and sets it chattering. News passes from neighbour to neighbour, from nerve cell to nerve cell: the beginning of vision.

How curious, that light should excite our eyes by turning *off* our photoreceptors. Not every eye works this way: invertebrate photoreceptors grow excited in the light. And if it's variety you're after, then there are all manner of light-detecting mechanisms – ones that have nothing to do with vision – that harness quite different chemical reactions. It was from these light-detecting mechanisms that vision evolved, for living things – with eyes or without, animals or vegetable – have always been superbly attuned to light.

Most living things get by without eyes and settle for a simpler, cheaper light sense. The 'phytochromes' in plants, which are sensitive to red and infra-red light, control how leaves and flowers fold and open and turn to the sun. They also control the plant's germination, growth and flowering. Many animals – including animals with well-developed image-forming eyes – use a generalised light sense to regulate their metabolic rhythms and trigger important behaviours. The burrowing sea anemone *Calamactis praelogus* bends towards the light. The flatworm *Convoluta roscoffensis* emerges from the sandy shallows at low tide so its symbiotic algae can photosynthesise. Pass a shadow across the sea urchin *Diadema setosum* and it will withdraw, waving its spines in a complex defensive pattern.

This simple movement towards or away from light has evolved, in higher animals, into an ambitious scheme of migrations and timed mating cycles. Salamanders and frogs navigate by a dermal light sense, suggesting that their skins are sensitive to polarised light. A generalised dermal light sense controls the flight rhythms and reproductive systems of birds. Pigeon chicks, hooded so that they cannot see, will still greet the light by wagging their heads and stretching their legs. The sea hare, a kind of sea slug, uses light to regulate its pattern of feeding and sleeping. However, its sensitivity to light resides not in its skin, but in its nerve fibres. Some animals, like leeches and starfish, have skins which are generally light-sensitive. Others, like the horseshoe crab, have a number of photosensitive sites which regulate their 'body clocks'. The more organised and centralised the animal's nervous system, the more likely it is that a general dermal light sense will be supplemented or replaced with a more evolved structure, located in or near the brain. This is the famous pineal body, a third eye present in all vertebrates, including humans. The first recorded discussions of the pineal organ in humans come from Hindu writings of around the sixteenth century, whose writers believed that the pineal was a form of eye. They were not wrong, though not until 1964 was it conclusively demonstrated that light penetrates the human brain. Until then there had been a tendency, among early evolutionists, to treat the human pineal 'gland' as a functionless left-over from our reptile past – as vestigial and uninteresting as the appendix. Its unassuming appearance no doubt played a part: this tiny, pine-cone shaped organ is only six millimetres long, and weighs just one-tenth of a gramme.

The pineal body's role in controlling the vertebrate body clock was discovered by a slightly roundabout route. In 1911 Karl von Frisch – who went on to win a Nobel prize for decoding the language of bees – was studying for his University Teaching Certificate at the University of Munich. Thanks in no small part to the example of his illustrious uncle, Sigmund Exner (the man who first unpicked the workings of the insect eye) Frisch was fascinated by how other animals perceive the world. In particular, he was curious to know how animals react to light. How, for instance, does the European minnow 'know' to adapt its coloration to remain camouflaged in shallow, sun-dappled water? To understand this mechanism, von Frisch shone a light on the head of a blinded European minnow. The minnow's skin, chameleon-like, began to pale – just as it would when a sighted minnow swam into sunlight. By isolating the light-sensitive region on the fish's head, von Frisch was able to show that the fish's changes in coloration were triggered by its pineal. Even if the fish were blinded, its pineal body still 'saw'.

The link between the activities of the pineal body and changes in skin coloration, though interesting, seemed to have no universal significance until, in 1958, Aaron Lerner isolated a substance from the pineal bodies of cows and studied its effects on known skin pigments. Because it caused granules of the pigment melanin to contract, he called his newly isolated pineal extract 'melatonin'.

In turn, the significance of melatonin's presence in the pineal organ was not fully understood until 1974, and came about largely thanks to the work of Julius Axelrod (1912–2004), an unpromising medical student who spent his early career in New York, testing vitamin supplements in milk.

'Successful scientists are generally recognised at a young age. They go to the best schools on scholarships, receive their postdoctoral training fellowships at prestigious laboratories, and publish early,' Axelrod wrote in 1988. 'None of this happened to me.'[1]

Axelrod had no illusions about his career. 'I expected that I would remain in the Laboratory of Industrial Hygiene for the rest of my working life. It was not a bad job, the work was moderately interesting, and the salary was adequate.' But he had not counted on the friendship and mentorship of Dr Bernard Brodie, a researcher at Goldwater Memorial Hospital. Encouraged by Brodie, Axelrod, though he had no formal qualifications, established himself in pharmacological research, studying analgesic drugs and the effects of caffeine and

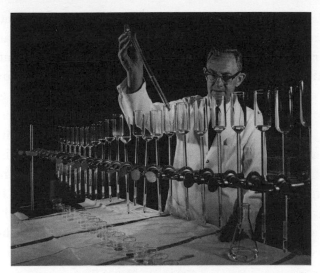

Julius Axelrod.

ephedrine. By the mid-Fifties, Axelrod was recognised as a pioneer in the study of drug metabolism and addiction, and was invited to set up his own laboratory at the National Institute of Mental Health. There, Axelrod and his co-workers began to unpick the chemical signals by which nerve cells communicate with each other – work which earned Axelrod a Nobel Prize in 1970.

Axelrod was drawn to the pineal because of the chemical provenance of that curious molecule melatonin. The nucleus of the melatonin molecule is serotonin, one of life's most useful building blocks. (It turns up in cephalopods and amphibians, in the pineals, retinas and associated retinal tissues of all vertebrates, in figs, plums, bananas . . .) Serotonin, because of its structural resemblance to LSD, was then believed to trigger psychoses. Axelrod found this assumption was incorrect. At the same time, he discovered that the pineal, far from being vestigial, was a biochemically active organ. But what was it for? With his colleagues Herbert Weissbach, Richard Wurtman and Sol Snyder, Axelrod revealed that the serotonin-melatonin cycle is a kind of biological clock, regulating both our daily cycles of sleep and metabolism and also the reproductive cycles of oestrus and menstruation. Further, the clock's multiple rhythms were synchronised by changes in light level. The pineal set itself by detecting daylight. It was, if not exactly an eye, at least a light detector of considerable sophistication.

The closer we look at them, the more similarities we find between the eye and the pineal. Architecturally they are quite different, but at the level of fine structure, there are some arresting parallels. The pineal photoreceptor cell is structurally similar to the cone cell. Genetic studies have since shown that cones are more ancient than rods; that rods evolved out of cones.[2] Might cones have evolved from pineal receptors? In a sense, it doesn't matter. Their similarity may simply be another, eloquent product of convergent evolution. The significant point is that pineal receptors and cones use virtually identical opsins, and the biochemical processes that follow their exposure to light are virtually interchangeable.[3]

The eye is not an isolated miracle. It is simply one particular – undeniably spectacular – species of light-sensing organ.

3 – *eyeless* in Basel

To suppose that the eye, with all its inimitable contrivances for adjusting the focus to different distances, for admitting different amounts of light, and for the correction of spherical and chromatic aberration, could have been formed by natural selection, seems, I freely confess, absurd in the highest possible degree.

So Charles Darwin begins his uphill task: persuading the readers of *The Origin of Species* that even 'organs of extreme perfection and complication' are not the product of a rational creator, but of a blind mechanical-historical process: evolution.

Why is Darwin so obviously girding himself for battle? Why is the eye's evolution so hard to credit?

Darwin expects his theory to be challenged on two fronts. For some, there does not seem to have been enough time for anything so perfect as the eye to have arisen by chance. Arguably the most elegantly presented statement of incredulity comes from the Reverend William Paley's book *Natural Theology* published over half a century earlier, in 1802: 'Were there no example in the world of contrivance except that of the eye, it would be alone sufficient to support the conclusion which we draw from it, as to the necessity of an intelligent Creator.'[4]

For others, it seems far-fetched that a patch of skin could ever evolve into an eye following the principles Darwin laid down. According to Darwin's theory, the present form of a species could only have arisen because every little intermediate stage of development conferred some survival benefit upon living individuals. But how can anything so complex as an eye have arisen by stages?

To these established sources of incredulity, we can now add a third. Consider the number of times the eye has arisen in nature (estimates wobble wildly between forty and sixty). Consider the range of different tissues, pigments and structures that go to make an eye, not to mention the range of behaviours that come packaged with the ability to see. When we find all these factors coming together in the same ways, not just once, but again and again, we cannot help but wonder whether even 538 million years gives evolution enough time for so many elaborate conjuring tricks.

No less dedicated an evolutionary biologist than Walter Gehring felt the need to voice this worry, in his book *The Homeobox Story*: 'Because the evolution of the prototype eye, at a stage before selection can exert its effect, must be a rare event, the independent evolution of so many prototypes represents a serious problem that is difficult to reconcile with Darwin's theory.'[5]

When an eye begins to distinguish images, the evolutionary pressure to see better than your neighbours will be colossal. But what is the evolutionary pressure on an 'eye' that is no more than a ribbon of pigment in a single cell? Gehring has no problem with the idea that the eye evolved. His problem is with the idea that up to sixty versions could have evolved independently. Even the most primitive eye is made of complex materials – materials that have their own evolutionary tale to tell. The coming-together of all these ingredients had to be, in Gehring's words, 'a rare event'. It was not driven by evolutionary pressures anything like as powerful as those which pit visual predator against visual predator in a race to develop keener and keener eyesight. So, Gehring believes that the eye arose only once, and all its many and varied versions are cousins.

* * *

As I write this, a housefly has just landed on the remains of the cake
I was eating. It's looking at the cake with immovable compound eyes,
which dominate its head. I am looking at it through squishy vertebrate
eyes that still carry the tang of ancient oceans. I can just about imagine
that the fly and I are related. We both have symmetrical body plans;
we both have heads, and mouths, and anuses. But there the similarities
end and a staggering list of differences begins. Even our skeletons bear
no resemblance: I keep mine on the inside, while the fly is encased in
articulated armour. Then there are our eyes. How can the bulbous,
faceted horrors of a fly's eye be related to my exquisite, camera-like
structures?

Flies, cats, dogs and people all start life as a single cell. The differences emerge over time, but at the start, we all look alike. Early in
development, there are basic structural similarities among all animals.
Heads, mouths, guts; if we have limbs, they extend from our trunks.

If the life we see around us today began in one place, at one time,
then living things must share a common ancestor. At the very earliest
stages of development, we resemble that ancestor, if not exactly, then
at least in the sense that we possess those features which we have
inherited, and which we all share. The German populariser of evolutionary theory, Ernst Haeckel (1834–1919), made much of the likely
similarity between an animal's foetal form and the shape of its ancestors.
Taken too literally, the idea – known by its (not very catchy) catchphrase
'ontogeny recapitulates phylogeny' – can lead the gullible paleontologist
down all sorts of garden paths. Still, it's a good rule of thumb.

Studying animals at the earliest stages of their development gives
us clues about human ancestry, and our relationship to other animals.
If animals came out differently each time, there would be very little
point in being an embryologist. If your dog had puppies, and one
puppy had four legs, one had six legs, and one had two legs growing
out of its head in place of its eyes, you would not think much of
those who spent their professional lives looking for a hidden order
in nature. At the other extreme, one would be unlikely to sign up
for a biological science course if humans were all the animal life the
planet had to offer; if humans bore humans exactly like the humans
before them, and the fossil record, however far back you searched,
contained layer upon layer upon layer of human remains. (In such

a world, foetuses would presumably look like miniature men and women. Even that first cell would be human-shaped, with little arm- and leg-like projections . . .)

Embryology is worth doing because in the real world, animals make *more or less* exact copies of themselves. Dogs make offspring that are like dogs, and people make people – most of the time. The flexibility that allows animals to change over time – so that several new species may arise from a common ancestor – needs there to be an outside chance that what emerges from the egg or the womb is not like its parents. Once in a blue moon, dogs do have six legs. Choctaw County Historical Museum in Alabama has one in a jar. And now and again, new-born humans turn out to be not very human.

Many women produce monsters: they just never know it. Human infants whose body plan is in disarray rarely make it out of the womb alive; really gross errors crop up so early in the development cycle that the fótus is spontaneously aborted long before the mother is aware of being pregnant. Over half of all pregnancies go disastrously wrong – we just never notice them.[6]

Much of the study of early development concerns what happens when the 'hidden order of nature' goes awry. Why are six-fingered children so rare – and why do they happen at all? By studying the exceptions one can, with a following wind, begin to work out the under- lying rules that govern an animal's development.

Walter Gehring claims that his career in molecular biology began the day his uncle sent him a mysterious box. Inside the box was a pile of brown, rather unprepossessing butterfly pupae. Young Walter, following the instructions in his uncle's letter, put the box up in the attic – and then forgot about it. The following spring, stumbling over the box again, he opened it. The butterflies had just emerged. Gehring stared in wonder: what had turned an ugly pupa into a beautiful butterfly? 'These mysteries,' says Gehring, 'have haunted me for my entire life.'

'With New Fly, Science Outdoes Hollywood', screamed the front page of *The New York Times*, some fifty years later, in 1995.[7] The *Times* was excited by experiments Gehring had been conducting on fruit flies. Gehring, in pursuit of the secrets of development, had managed

Walter Gehring in 1964.

to grow eyes on flies' knees, in place of antennae, and even on their wings. The popular response to the news was as boisterous as it was predictable. Victor Frankenstein, it seemed, was alive and well and living in Basel. One of Gehring's own photographs, showing a fly with extra eyes, was used on a poster in a campaign opposing gene technology.

This was a bit rich. For a start, Gehring's experiments were not strictly news. In 1935 a former bookseller, Hans Spemann, won a Nobel prize for getting eyes to grow on the bellies of amphibia. And the only reason either man was able to accomplish such gruesome feats of magic is because nature itself constantly throws up sports of this sort. One of the photographs in Richard Dawkins's splendid account of evolution, *Climbing Mount Improbable*, is of a frog whose eyes have grown, not out from the top of its skull, but down, into the roof of its mouth; it was discovered, hale and hearty, in an Ontario garden.

Thirty years after the publication of Darwin's *The Origin of Species*, the embryologist William Bateson gave a name to the mechanism by which animals acquire one organ or extremity in place of another: *homeosis*, the change of something into the likeness of something else. (Bateson was fond of coining neologisms. 'Genetics' is another of his.) Homeosis can express itself either as a birth defect, or during repair to a damaged or missing organ. When a lobster loses an antenna, another antenna grows in its place. After more than a couple of amputations, the luckless crustacean loses the plot, and grows a leg there instead.

Spemann's fervent hope was that, by understanding homeotic mutation, he would come to understand what an animal's body plan (or 'organiser centre') looked like and what it was made of. Gehring's first experiments – which bore more than a passing resemblance to those of Spemann – revealed that fruit fly embryos are arranged very simply.

Gehring's descriptions read more like the patents for toys than studies of real animals. The embryonic fruit fly is a set of discs, called 'imaginal discs', because each carries a genetic blueprint of what it wants to be. If the top disc wants to be a head, you can guarantee that the bottom disc wants to be a thorax. Rearrange the discs, and you interfere in predictable ways with the body plan of the fly.

From these experiments, Gehring was able to isolate the genes responsible for giving each disc its destiny. These ancient genes didn't directly *make* anything you could point to. Rather, they turned on other genes, which turned on others, thus shaping the development of large, ordered structures; a cascade of ever more numerous, ever more specific instructions, triggered by one broad 'executive' order.

When not given witty monikers by whimsical geneticists (the 'Harry Potter' puberty gene, anyone?) genes are named after the conditions and abnormalities that led to their discovery. This, inevitably, means that their name commemorates what happens when they *don't* work. The gene that triggers the development of eyes in fruit flies – discovered in 1915 – is called '*eyeless*'. Although Gehring experimented with *eyeless*, he had no particular interest in eyes, and it was a fluke finding by a junior colleague that triggered the work for which he is best known.

In 1993 Rebecca Quiring was hunting for the body-plan genes of fruit flies – the ancient genes which tell the imaginal discs of the foetal fly what form to take. On the assumption that certain proteins, called transcription factors, bind selectively to specific genes, Quiring cloned a known protein to use as a 'lure'. But the protein insisted on binding only to a gene – *eyeless* – that they already knew about. It seemed a pity for Quiring to just throw away all her work, so Gehring suggested that she submit the genetic sequence of her protein to the European Molecular Biology Laboratory's database in Heidelberg. If she had inadvertently cloned *eyeless*, well, that was a result of sorts, and it could at least count towards her PhD.

The acres of paper the Heidelberg computer spewed out in reply to Quiring's submission suggested nothing so much as a printer glitch. It took a little while to figure out what this mass of data actually was – and when Quiring did, she skipped down the corridors in high glee. Certainly, her protein corresponded to the fruit fly's *eyeless* gene. But the Heidelberg computer, not content with making one match, had found others: *smalleye,* a gene from a mouse; and *aniridia,* a gene from a human being. All three are essential to the proper development of eyes. A mouse with a missing *smalleye* gene is born with severe facial disfigurements, including undeveloped eyes and sealed nostrils. Humans with a missing *aniridia* gene have missing or malformed irises, cataracts, glaucoma, squint, and (suggestively, given the role of *smalleye* in mice) a loss of the sense of smell.

Were *eyeless, smalleye* and *aniridia* just different names for the same gene? It was not hard to believe that mice and humans might share a gene for eye development. Men and mice are both mammals, and have a fairly recent common ancestor. What really stretched credulity was that the gene responsible for mammal eyes is also responsible for the compound eyes of fruit flies. The common ancestor of flies and humans lived about five hundred million years ago. Snails are more closely related to flies than we are. Could a single control gene really trigger the development of such different eyes as the faceted compound eyes of a fly and the wet cameras of the human being?

Gehring's team had to demonstrate that the *eyeless* gene was the one gene ultimately responsible for the development of the fruit fly's compound eyes. So Gehring returned to the work he had conducted thirty years before on the 'imaginal discs' that make up a fruit fly embryo. By turning on the *eyeless* gene in different imaginal discs, the team induced ectopic eyes – eyes that grew where they did not belong: on wings, and knees, and in place of antennas. One fly had fourteen extra eyes which, although it could not see through them – they were not connected to the brain – were sensitive to light.

What would happen if they replaced the fly's native *eyeless* gene with *smalleye?*

Since that first burst of print-out from Heidelberg, a slew of *eyeless*-like genes in other species had come to the department's notice, including genes in squid, sea-squirts and zebra fish. All were essentially the same, with a little local variation. (This is why – to run ahead of ourselves for a second – modern accounts tend to drop the species-specific names and lump all these genes under one convenient, if unmemorable, label: *Pax6.*) So, despite *The New York Times,* no one seriously expected anything freakish. (A mouse eye on the head of a fly; quivering nostrils in place of mouthparts . . .) *Smalleye* and *eyeless* are about ninety-four per cent identical.

On the other hand, no one could be sure how well *smalleye* would stand in for its insect cousin. Early results were very unpromising. Then, the day before they planned to give up, Gehring's collaborators, Georg Halder and Patrick Callaerts, found the first spots of red pigment on a fly's legs. A week later, the first facets of an insect eye had begun to form. Soon the flies boasted spectacular working eyes on their legs, their antennae, and their wings. The mouse gene *smalleye* had directed the creation of perfect compound eyes.

In their paper in *Science* in March 1995, Gehring, Halder and Callaerts declared: 'The observation that mammals and insects, which have evolved separately for more than 500 million years, share the same master control gene for eye morphogenesis indicates that the genetic control mechanisms of development are much more universal than anticipated.'[8]

A year later, Gehring went much further when, in response to a critical letter in *Science*, he announced that his work went 'against the dogma of eye evolution that can be found in most textbooks.' This claim – and in particular the use of the word dogma – was, frankly, inflammatory. Gehring must have thought attack was the best form of defence, for his target could not have been more highly esteemed. Ernst Mayr (1904–2005), considered by many the founding father of evolutionary biology, was the man who had first come up with the idea that the eye had evolved independently many times. According to Gehring's letter, Mayr's notion was the new orthodoxy – and he was the man to demolish it.

Setting aside its choice of language, Gehring's grandiose claim is founded on the rather dangerous assumption that he has found the gene 'responsible' for eye development. At first glance this seems undeniable – Gehring's ability to induce ectopic eyes in fruit flies using a *smalleye* gene from a mouse demonstrates as much.

But what does it actually mean when we say that a gene is 'responsible' for something? Genes are not an army of microscopic labourers. They don't build organised structures, the way construction workers erect buildings. Genes squat inside cells. Genes that are turned off do nothing; those that are turned on control the production of different proteins.

And that's all. An animal's life can be seen as a 'working out' of the billions of little directives issued by genes in every one of the animal's cells, but these directives are dependent on circumstances. Every cell contains instructions on how to build a head but, thanks to induction (the 'dance of development' in which different foetal tissues touch and communicate chemically with each other) the instruction 'build a head' is heeded only once.

The point is well made in the very letter to *Science* that elicited such a strong reaction from Gehring. W Joe Dickinson and Jon Seger, from the University of Utah, point out that the *eyeless* gene is turned on in plenty of cells that have nothing to do with the eye. Indeed, if the gene is turned on in all cells, 'this does not convert the entire embryo into eye structures.' Dickinson and Seger conclude: '. . . we take this to imply that [*eyeless*] originally served some basic developmental process other than eye induction.'[9]

Many a squid would agree: the squid *Pax6* gene controls the development of its tentacles, suggesting that the gene has a role in the formation of many different kinds of sensory organ. Certain species of blind worm use a version of *eyeless* to control the shape they acquire as they grow. Every living thing has a genetic vocabulary to write in. As Dickinson and Seger's letter reminded the readers of *Science*, a gene called *hedgehog* crops up in the wing development of flies and birds – structures as markedly different in their way as insect and mouse eyes. This does not mean that the fly's wing and the bird's wing are functionally 'related'.

'What we claimed,' said Mayr, 'and it's as correct as ever, is that the eyes themselves, the photoreceptive organs, developed independently at least forty times. This is not in the slightest touched by finding that there are genes used in making eyes that existed long before eyes. You should go to the species that have no eyes but have this gene and find out what it's doing.'[10]

Genetic studies like Gehring's have revealed a rich pre-history for the eye, without which its emergence and re-emergence, forty times over, would be inexplicable. They tell us what tools were used to build the first eye, and roughly when those tools first became available.

But they cannot tell us when or how the first eye arose, or which animals first saw.

4 – The shoeshiner's tale

. . . man in thinking greatly errs particularly when inquiring after cause and effect; the two together constitute the indissoluble phenomenon.

J W von Goethe[11]

Louis Jacques-Mande Daguerre was born in a small town on the outskirts of Paris in 1787. He did not invent photography. In the mid-nineteenth century, there were many competing processes for recording and fixing photographic images, and historians of photography salute many pioneers. The architect, stage designer and illusionist Daguerre has the distinction of being the man who sold photography to the French government. The government then freely gave the process to the world, but other, better processes were already abroad, and the gesture fell flat.

Photographs taken by Daguerre's method – daguerrotypes – are the earliest photographs we have, because Daguerre cracked the vexing problem of how to stop the images from fading. His other big achievement was to reduce the photographic exposure time from some eight hours to only thirty minutes. This improvement made it possible to take photography out of the laboratory and capture reasonably faithful real-world scenes.

His 1838 daguerreotype of the boulevard du Temple, for instance, required an exposure time of only ten or twenty minutes and, consequently, the play of light and shadows on the trees and buildings is remarkably life-like. We can tell, at a glance, that this photograph was taken in the middle of the day. But where are the people? The street, one of the busiest in Paris, must have been heaving with them. But where are they?

With an exposure time of tens of minutes, Daguerre knew he would not be able to capture the human business of this busy Parisian street. Indeed, the novelty of the image – an empty boulevard, indeed! – was probably uppermost in his mind when he framed the shot.

But human life is not altogether excluded. In the bottom left of the picture is a shadowy figure – a man having his boots polished.

As a German magazine of 1839 observed, he 'must have held himself extremely still for he can be very clearly seen, in contrast to the shoeshine man, whose ceaseless movement causes him to appear completely blurred and imprecise.' Imagine we had no newspaper accounts, no diaries, no histories to work from. Imagine we only had Daguerre's photograph. What would it tell us about the boulevard du Temple? What would we conclude from our study of this snapshot in time? Would it be unreasonable to conclude that Paris had been virtually abandoned, haunted only by a handful of tall gentlemen and wraith-like bootblacks?

Fossils are evolution's daguerrotypes, recording – in their peculiar, mysterious, and eccentric ways – the progress of life on this planet. They are our most valuable source of information about life's history; and yet, very often, they lead us horribly astray.

In a letter to an American colleague, Charles Darwin famously declared that thinking about the evolution of the eye gave him a 'cold shudder'. These occasional concessions to incredulity are a favourite motif in Darwin's writing (elsewhere, he says that thinking about the tails of peacocks made him physically sick). He uses this piece from his rhetorical armoury when he introduces readers of *The Origin of Species* to evolution – a process of nature which is 'insuperable to our imagination' because of its sheer scale, ubiquity and (in most cases) extreme slowness.

No one looking at a countryside scene can 'picture' evolution in action, any more than anyone, looking at the star-lit sky, can 'picture' the scale of the galaxy. Darwin took biology out of the anthropomorphic world of mythical and biblical parable and into the astronomically scaled world of evolutionary time. We should expect a shudder or two.

What really had Charles Darwin on the run was the lack of fossil evidence.

For evolution to be truly a force in nature, it had to have given rise to all life, developing it by miniscule increments. In Darwin's day, however, there was no evidence of the existence of life earlier than the Cambrian period. Animals had simply appeared around 540 million years ago (or 540Ma, to use the scientific shorthand). Even more worrying for Darwin and his supporters, these first life forms were not remotely primitive. They had claws, jaws, teeth, tentacles – and eyes.

Older rocks had been identified and studied. Earlier ages had been mapped by geologists. But none of these older rocks contained fossils. On the evidence of the rocks, none of these earlier times had boasted life. It was as though, far from emerging slowly from microscopic animalcules and slime, life had simply exploded 543Ma, fully formed and ready for action.

For many of Darwin's generation, the evidence for the instantaneous appearance of life on Earth during the Cambrian period offered the possibility of a reconciliation between evolution and divine creation. The most articulate stab at a reconciliation came from William Buckland, professor of geology at Oxford. Some twenty years before Darwin's *The Origin of Species* was published, Buckland pointed to the Cambrian Explosion as a possible, though distinctly un-Biblical, creation event.

Buckland's opposition to Darwin's theory of evolution seems ironic now. It was he who wrote up the first description of a dinosaur fossil, the 'Great Fossil Lizard of Stonesfield'; he who found evidence of mammals living in the 'age of reptiles'; he whose grasp of the principle of 'survival of the fittest' made for some memorable lectures. (One of his students recalled how Buckland '. . . rushed, skull in hand, at the first undergraduate on the front bench – and shouted, "What rules the world?" The youth, terrified, threw himself against the next back seat, and answered not a word. He rushed then on me, pointing the hyena full in my face – "What rules the world?" "Haven't an idea," I answered. "The stomach, sir," he cried.')

Buckland's opposition to evolution was not willful or blinkered, and at the time, the weight of evidence lay on his side.

Much of the fossil record of life on Earth remained to be discovered, as both Buckland and Darwin knew perfectly well. Even the terminology they used was tentative and uncertain, which is why Darwin never writes about the Cambrian period itself, but about the 'Silurian', a word that has since shifted its meaning to refer to a completely different slice of geological time. Writing of the sudden appearance of the complex, woodlouse-like trilobites in the fossil record, Darwin had no doubt that 'all the Silurian trilobites have descended from some one crustacean which must have lived long before the Silurian.'

Neither Darwin, Buckland nor any of the early champions and opponents of his disquieting theory lived to see this expectation confirmed. During the second half of the nineteenth century there were scattered reports of fossil finds in pre-Cambrian strata. But Darwin's theories came to be accepted, not so much because of the fossil evidence, but primarily because they were able to explain circumstantially the diversity of modern life.

Not until 1946 did an Australian mining geologist happen upon the first well-preserved pre-Cambrian fossils. Exploring the Ediacara Hills, a range of mountains north of the city of Adelaide, the mining geologist Reginald C Sprigg found fossils of what looked like soft-bodied organisms. Some resembled jellyfish; others were more like shellfish, or worms.

Whether or not the Ediacaran faunas were an odd, non-animal 'trial run' at multicellular life or an early, eyeless and toothless period of animal evolution, their existence has done little to erode the mystery of the Cambrian explosion. The plain fact is that within five million years, between 543 and 538Ma, life on earth transformed completely. The way animals were organised, how they lived and how they behaved, underwent a massive sea-change, from blind drifting and casual grazing to visually guided predation, defence, camouflage and evasion. It all happened so quickly that the fossil record – laid down excruciatingly slowly over millennia – can no more record its progress than Louis Daguerre's camera, pointed down the boulevard du Temple in 1838, could have captured its busy traffic of carriages and people.

What lessons should we draw from the Cambrian explosion? For the American paleontologist Steven Jay Gould (1941–2002), writing in the late 1970s, the sudden appearance of so many odd and arresting animals suggested 'a unique time of organic flexibility, before major developmental pathways became irrevocably set'. Maybe animals didn't evolve slowly at all. Maybe they evolved intermittently, quickly, and wildly.

The theory of 'punctuated equilibrium', promulgated by Gould and his colleague Niles Eldredge, implied an as-yet-unidentified genetic mechanism capable of triggering a Mardi Gras of novel forms.

Genetics, however, conspicuously failed to throw up any evidence of this mechanism.

The first animals appeared about 750Ma ago. We can be fairly sure of this, because we have a good idea of how often complex proteins such as haemoglobin mutate. If we count up the number of genetic differences between the haemoglobin used by one species and the haemoglobin used by another, we can estimate the age of their common ancestor. (This neat trick was inspired by George Wald's work on rhodopsin.) By counting up the differences between the most unrelated species, we arrive at the age of the oldest-ever animal – the common ancestor of all animals everywhere – and we find that it lived 750Ma. Gould's conjecture thus finds itself in a heap of trouble, because 750Ma is two hundred million years *before* animals strode into view in the Cambrian.

Since the kinds of genetic innovation that made animals possible occured two hundred million years before the Cambrian explosion, they can hardly be said to have 'caused' it. The only genetic event we know of that might have triggered an event as extraordinary as the Cambrian explosion was the first appearance of Gehring's body-plan genes – *Pax6* and the like. (One imagines these newly-minted genes, like demented Christmas elves, stringing together segmented novelties like toy trains and tossing them into warm Cambrian seas.) But as Gehring's work has shown, *Pax6* and its fellows are shared by species as profoundly unrelated as humans and houseflies. These genes are therefore as old as our oldest common ancestor – too old to have triggered the explosion.

Once Gehring's body-plan genes had been shown to pre-date the Cambrian, Gould's conjecture was left high and dry – a soft target for its critics.[12]

If there is no genetic mechanism to explain the Cambrian explosion, there is no shortage of other plausible explanations. Perhaps an extinction event cleared the way for novel forms of life. The Earth has experienced at least six major extinction events – from asteroid strikes to plagues of clever apes. The 'ape plague' story is still being written, but the outcome of the other five events was always the same: a period of rapid evolutionary development among the species that survived. The trouble is, the Cambrian explosion was not preceded by an extinction event. In some areas, it is true, the pre-Cambrian fauna (called Ediacaran, after the hills in which their fossils were first discovered) died out before the Cambrian fauna took over, but in other places, Ediacaran life seems to have been gently edged out by Cambrian newcomers over millions of years. An increase in oxygen levels, around 545Ma, coincided with the Cambrian event, but oxygen levels have rocketed and plunged on numerous occasions during Earth's history, with surprisingly little effect on the biosphere. Maybe there were more shallow seas than before, as sea levels rose, drowning the Cambrian continent's barren lowlands (there was, of course, no life on land at this time – or no life to speak of; an algal colony here or there, perhaps). But why should an increase in shallow seas make everything grow so big, and acquire such novel forms?

These are all good ideas, and there are plenty more. But none of them is complete or compelling.

Part of the problem is to do with the sort of explanation we are looking for. It is surprisingly difficult to talk about what 'caused' the Cambrian explosion, because every cause is the effect of a preceding cause. Where do we draw the line?

Three things we can be certain of about the Cambrian explosion: animals got big, they got hard (hard enough to be preserved in the fossil record), and they got clever.

We do not know why animals grew bigger in the Cambrian, but we do know that getting bigger isn't hard to do. When the dinosaurs died, mammals blew up like balloons to fill the suddenly empty niches. Increasing size may explain why Cambrian animals first began to develop hard parts. Hard surfaces provide purchase for muscles, so that movement becomes more powerful, directional and coherent. The so-called 'small shelly fauna' that are the first Cambrian fossils are sometimes the skeletal remains of whole creatures, and sometimes the individual leaves of a 'chain mail' covering much larger animals. The use of the word 'chain mail' is apposite, because alongside evidence of hard parts comes evidence of predation: some of these shells have holes bored through them.

The Cambrian was a violent time. The fossilised eggs of bristle worms and jellyfish contain foetuses so well developed, they are almost adult; an expensive reproductive strategy, employed only when newly-hatched young have to be ready for everything the world has to throw at them.

What was out there? For a start, there was *Anomalocaris*, a metre-long stalk-eyed nightmare with a mouth like a camera iris. We know it ate animals, because they are preserved inside *Anomalocaris*'s fossilised guts. Then there were the trilobites. These came in all shapes and sizes and they were around for a very long time: about 300 million years. Some trilobites were bottom feeders, some were fed by bacterial mats living in their gills, and some were hunters. The very first trilobite species we know of had eyes. And not blurry squitty eyes, but gorgeous, faceted, compound eyes, exquisitely constructed and assembled to a design that has never been repeated, for all that eyes have arisen at least forty times on Earth.

With size and armour came cleverness. We can measure an animal's cleverness by studying its tracks. The fossilised tracks of Ediacaran animals are touchingly simple; the tracks of complacent grazers, little fleshy lawnmowers, trimming back the algal mats of primaeval seas. There are burrows, but they are shallow and often drilled horizontally into a sloping bank, and their walls are neither lined nor armoured, offering little in the way of defence. Indeed, there seems to be nothing to defend against. If there were predators, then they were opportunists, incapacitating or killing whatever hapless creature happened to cross their path. This lackadaisical style of predation could not have inspired the evasion strategies we find among animals today. Speed, strength, size, hardened shelters, and hard skeletons are all excellent defences against predation, but there is no evidence for any of them among the Ediacaran fauna.

Cambrian tracks tell a very different story. The surface is perforated with vertical burrows drilled into the sediment. Monstrous tracks weave back and forth across the sea bed. Pits in the rock mark the frantic excavations of rapacious hunters. These are not the tracks of harmless, fearless grazers – these are the spoor of conflicts, battles, competitions for resources; of the desperate measures animals take to eat and not be eaten.

What triggered these extraordinary physical and behavioral novelties?

When animal life consisted solely of tiny, millimetre-long worms and planktonic drifters, there was no need for vision, for there was no animal big enough to see. Visible light (including the ultraviolet) varies in wavelength between .35 and .8 microns – just less than the width of the smaller bacteria. Such animals are big enough to block the light – but only by a whisker. *Euglena gracilis* is 50 microns long. For *Euglena*, there can be no vision. Light detection, certainly. But not vision. Vision requires more than an occasional nudge from the crest of a light-wave. It requires a deluge, heavy enough to reveal patterns of shine and shade. *Euglena* is only sixty times larger than the wavelength of red light; it is just too small.

Just as there is a minimum size for a true, image-forming eye, there is a minimum level of complexity. Vision requires a nervous system sufficiently complex to harness the dance of light within the eye. Why animals should have acquired such complexity is still a mystery. If an animal had a rudimentary light sense, perhaps a increase in size would have led directly to an increase in nervous complexity. As the eye grew bigger, and its light-gathering capacity increased, vision would have improved, maturing, over generations, from mere light detection into an ever-more suggestive view of the world. Provided with this extra information, the nervous system connected to this proto-eye would have evolved a growing sensitivity to rapid changes in light level, suggestive of the approach of a mate, or a predator. No one knows exactly why animals grew bigger in the Cambrian, but competition triggered by the evolution of vision would certainly have been influential enough to drive an increase in body size.

By the same token, we do not know why animals chose, 543Ma, to acquire teeth, legs, grasping forelimbs, evasion tactics, mining skills, colours, bristles, shells, mirrors, and all the other paraphernalia of modern life, but we do know the reason they acquired them. Like Adam and Eve caught naked and defenceless in the Garden, Life became aware, for the very first time, that it was being watched.

There's no point having legs if you can't see where you're going. (The other senses will give you some idea of your environment, but only vision will give you the instantaneous information that makes agile movements possible.) Teeth and armour, too, are the product of an arms race that could not have got started without vision. When the chances of getting snared by a predator are little more than random, there's no reason to evolve heavy, expensive defences against misfortune. But once the predator can see you, and pursue you, the chances of falling victim to its attentions are astronomically increased and a suit of armour becomes a worthwhile investment.

One alternative is simple inedibility. No doubt, Ediacaran animals employed venoms and toxins for their personal protection. They are cheap and easy to make, and just the sort of budget weapon a soft-bodied animal could afford. But faced with animals who prepare their food first and complain to the chef later, like a metre-long *Anomalocaris*, or a trilobite, which although only a few centimetres long, has spiny legs that double up as meat shredders, toxins alone are not enough. What our prey animal needs is spines. Lots of spines. *Marella splendens*, a large (two centimetre-long) blind bottom feeder is so spiny, it resembles a stripped skeleton more than a living animal. It was the Volvo of the Cambrian shallows, tootling complacently about in its safety cage. Other creatures were more direct, growing hedgehog-like spines, and even blades. Many trilobites arc fantastically spiny; others were able to roll up into an armoured ball so exquisitely fashioned that, in some cases, it came with a safety lock; the more you tried to prise the thing open from the outside, the more secure the fastening became.

Armour is a last-ditch defence; it would bc far better to go about one's business without interference. To avoid a blind predator, one need only keep quiet, move slowly and try not to be smelly. To avoid a sighted predator, one has to do something, and the most obvious thing to do is hide. There is a type of Cambrian sandstone nicknamed 'pipe rock' by geologists because of the number of vertical burrows puncturing it. It looks like fossilised crumpet.

Burrowing has its limitations. If you spend all day hiding in your hole, where is your food going to come from? The birth of vision heralded ever-more-subtle strategies for concealment; ways of hiding that did not involve burrows, and permitted animals to move – however cautiously – through their environment. Animals evolved to disguise their own nature; to appear inedible, or dangerous, too large or too awkward to swallow. Some hid the fact that they were food at all. Camouflage, warning colours, mimicry, deceit, or complex behaviour could, of course, also be harnessed by predators. They too could disguise themselves, so that their prey was lured towards them. Since most animals were both prey and predator, quite sophisticated and subtle disguises were bound to evolve. The sheer complexity of the Cambrian continues to reveal itself, and the most startling recent finds, at the time of writing, are those reported by Andrew Parker, a zoologist whose interest in the fossil record grew out of his studies of structural colour.

Structural colour occurs when a lighted surface throws off multiple reflections. These reflections are generated so close to each other that the light waves interfere; some wavelengths are completely damped, while others shine with redoubled force. There are two ways of generating structural colour: either light is reflected off a corrugated surface whose corrugations are fine enough to generate interference effects (such as a peacock's feather or a compact disc), or light passes through multiple translucent layers, each of which reflects a small portion of the total incoming light (like a fly's wing or a roll of clingfilm).

A well-organised surface texture of a certain fineness is likely to generate structural colour, so we should not be surprised to find them in nature. Yet we need to explain why animals should openly exhibit such colours. Pilots flying over the Brazilian rainforest have reported seeing brief flashes as blue morpho butterflies flap their beautiful, structurally blue wings above the forest canopy. Any animal sending out such a beacon of colour – brighter by far than any pigment – must do so out of some pressing need.

Structural colours make excellent signalling devices – something Parker discovered for himself in 1995 when, in Watson's Bay in Sydney Harbour, he caught two seed shrimps in courtship. The idea that seed shrimps have a courtship ritual at all seemed, at the time, a little far-fetched. These humble crustaceans are tiny, ranging from just a hundred microns to a few millimetres in diameter. And yet, as Parker watched, the male flashed a dazzling blue light. Seconds later, the pair were mating.

This extraordinary discovery led Parker on a scientific treasure hunt that ended among the fossils of the Burgess Shale in the Canadian Rockies, the world's richest and most celebrated Cambrian fossil bed, discovered by Charles Walcott in 1907. From the sunny shallows of Watson's Bay to the 538-million year old shallows of the Cambrian seems quite a leap – but the humble seed shrimp had covered most of the distance. Seed shrimps have been around for about 250 million years. Though it is possible that the seed shrimp acquired its signalling apparatus recently, it is equally possible – even likely, given how little the animal had changed over millions of years – that seed shrimps have been flashing seductive lights at each other since their inception. Parker was on the trail of the earliest colours in nature. Pigments, alas, do not survive the fossilization process. But structural colours do. It had already been established that lamp shells, living 350Ma, boasted multi-layer reflectors as iridescent as those of a butterfly's wing.

With the brashness that is his trademark, Parker looked for equivalent structures on some of the best-known fossils of the Burgess Shale: *Wiwaxia corrugata* (imagine an animated knife-block with the knives put in blade-uppermost); *Canadia spinosa* (much like a modern bristle-worm); and the walking safety-cage *Marella splendens*. To his delight, Parker was able to show that all three animals boasted structural colours. They were not merely colourful – they were shiny.

Why would a potential meal like *Canadia* need to shine? Parker argues that it is inconceivable that these structural colours were an accident. Animals this conspicuous would have perished in an eye-blink, were their colours not some form of generally understood warning. Were Cambrian predators sophisticated enough to be disconcerted by flashing lights? Were they capable of learning that sparkly creatures tend to stick in the craw?

Parker's discovery of colour in the Cambrian suggests a dangerous, fast-paced, colourful environment. It also suggests that Cambrian animals possessed a modicum of cleverness; the shape, texture and colour of things *meant* something to them.

According to Parker's recent book *In the Blink of an Eye*, the advent of vision powered the Cambrian explosion.[13] He even points to a specific eye as the culprit – that of a species of predatory trilobite. This is, to say the least, unfortunate: the earliest trilobites don't appear in the fossil record until three and a half million years after the explosion they are meant to trigger.[14] Otherwise, Parker's general idea – that the advent of vision triggered a massive explosion of species 543Ma – seems self-evident, not to say fatuous.

For an account of how the arrival of vision triggered natural variety, we do not need Parker's account, enjoyable as it is. We can turn to the ecological writings of Canadian atheist Grant Allen, writing in 1879:

> Insects produce flowers. Flowers produce the colour-sense in insects. The colour-sense produces a taste for colour. The taste for colour produces butterflies and brilliant beetles. Birds and mammals produce fruits. Fruits produce a taste for colour in birds and mammals. The taste for colour produces the external hues of hummingbirds, parrots, and monkeys. Man's frugivorous ancestry produces in him a similar taste; and that taste produces the various final results of the chromatic arts.[15]

Visual predation did not 'cause' the Cambrian explosion; visual predation, and the adaptational pressure it placed upon living things, *was* the Cambrian explosion. The question that needs answering – the question nobody has the answer to, and the one which inspired Stephen Gould's brave conjecture about a mysterious 'genetic event' – is what made sophisticated nervous systems possible in the first

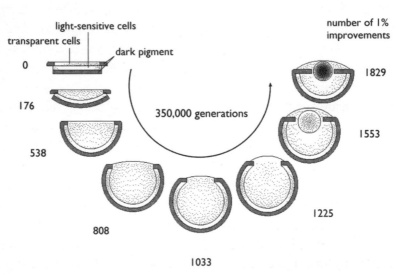

The results of Dan-Eric Nilsson and Susanne Pelger's seminal 1994 computer model of eye evolution. Their fish-like eye would evolve too fast for the process to be reliably recorded in the fossil record.

place? Until animals got big and complex, there would have been no acute, image-forming eyes, for no animal would have been capable of seeing.

Given that small genetic changes can lead to large alterations of physique, it may be that very little time indeed is required for a small, slug-like Ediacaran animal to develop a really quite startling Cambrian motley, complete with legs, claws, teeth, a suit of many colours, eyes and a complex nervous system.

How long does it take for at least the outward form of an eye to evolve? Two Swedish scientists, Susanne Pelger and Dan-Eric Nilsson, made a creditable first stab at the problem in 1994, designing a computer simulation of how a fish's eye could evolve from a flat, photosensitive surface. Even in the absence of any knowledge about the mutation rates and life cycles of real fish, this experiment was still useful, because it showed how a really good eye might evolve from a really bad one, smoothly, without awkward or useless intervening stages.

Almost every book on my research shelf trots out the famous diagram of their findings – reproduced here, yet again, with numbing inevitability. But its elegance earns it another outing. With startling visual clarity, it reveals how an eye may evolve, and how seemingly complex structures like a lens can arise from small, graduated changes in tissue density.

The rules Nilsson and Pelger set for their model were impressively few. They made one central assumption – that good eyes are better than bad ones – and otherwise let their model develop as it would. Each new generation, one part of the 'eye' was permitted randomly to alter its refractive index – the degree to which it bent light – by one per cent. The model was also allowed random changes of shape, again by small increments. If the new version was optically superior to its 'parent', it was permitted to generate 'children'. If optically inferior, it 'died'.

The result was decisive. Starting with a flat eye spot, the model deformed to form a shallow pit, marginally improving the eye spot's acuity. The dimple deepened until it was as deep as it was wide, like the eye of a flatworm. Then the edges of the pit began to close over to form an aperture.

The amount of available light dictates how narrow the aperture of this 'dimple eye' will be. The narrower the hole, the better the whole structure will work, casting an inverted image on the back wall of the eye. If the hole is too narrow, very little light will enter and the image, though sharp, will be too dim to see. The only way to improve an eye of this sort is to develop a lens. The stuff of which the lens is made is, in most cases, already present. Some natural 'pit' eyes are empty; the nautilus has an eye of this sort. But Nilsson and Pelger's eye very quickly came to resemble the eye of a snail, whose pit eyes are full of a more or less homogenous jelly. (It is not hard to imagine how this jelly is formed, since all sensitive tissues that are exposed to the elements have some form of mucosal protection.) Eventually, random changes in refractive index through the jelly created a perfectly round optical lens.

Of course, in that word 'eventually' lurk all manner of hard questions. Having shown that an eye could evolve, were Nilsson and Pelger able to say how long it would take to evolve? Indeed they were. Happily we know something about mutation rates, about how reliably these mutations will show up in individuals, and about how much individual variation there is in the population. By picking extremely pessimistic values for these and other variables, Nilsson and Pelger calculated how many generations it would require for their fish to stumble upon the 1,829 single-percentage improvements it takes to make an eye. The answer stunned everyone: 400,000 generations.

Assume our fish breeds once a year (another pessimistic estimate). That's fewer than half a million years for an eye to evolve from an eyespot. How would such an event be reflected in the geological record?

It wouldn't be reflected at all.

400,000 years is no time at all, geologically speaking. Even if we were, by some miracle, to find the needle in the haystack, the half-way stage of our fish's ocular evolution, how would we possibly know that was what it was? The rocks won't tell us that one fish lived a few hundred thousand years earlier than another. Geological time just can't be measured with such fineness. Our new-found fossil might simply be another species of fish, with less sophisticated eyes.

Computers were really not all that good in 1995. Ten years on, we could write much better models of all sorts of aspects of eye evolution, from the biochemistry and architecture of photocells to the stitching of retinas. But none of this work is likely to have the same headline-grabbing kudos as Nilsson and Pelger's original paper, which captures the dramatic evolution of whole eyeballs.

Nilsson and Pelger's model, like all models, depends on certain assumptions; not least, that before vision developed, patches of an animal's skin were – or could become – photosensitive. It assumes, in other words, a prehistory for eyesight: a period, who knows how long, in which animals acquired the neural sophistication to respond to a new-fangled sense. That prehistory has yet to be written. Perhaps it, too, is the story of a few eye-blinks of geological time. Or it might be an excruciatingly casual tale, taking two hundred million years or so between the emergence of animals and the eventual, tardy, but undeniably dramatic acquisition of image-forming eyes in the Cambrian.

Animals did perfectly well without eyes for two hundred million years. They ate algae. They slumped. They pulsed. But then there were eyes – and the fun began.

FOUR

The adaptable eye

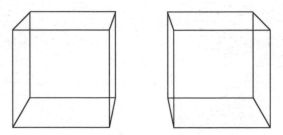

1 – 'A basically disastrous design'

'The apt,' writes Edgar Rice Burroughs, in his science fantasy *Warlord of Mars*, 'was our most consistent and dangerous foe.'

> It is a huge, white-furred creature with six limbs . . . Its two huge eyes inspired my greatest curiosity. They extend in two vast, oval patches from the center of the top of the cranium down either side of the head to below the roots of the horns, so that these weapons really grow out from the lower part of the eyes, which are composed of several thousand ocelli each.

Insect eyes – huge, unblinking, impersonal – are one of the more visceral and chilling pieces of visual repertoire available to writers of science fiction and the makers of horror films. And no wonder: our own eyes, with their bright whites, rainbow irises, responsive pupils, brows and lashes, not to mention their wide racial variation, have evolved to communicate and carry meaning. Our

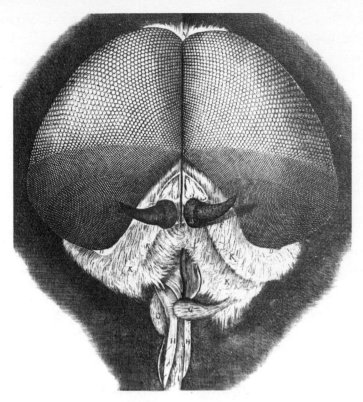

Robert Hooke's startling portrait of a fly is the first high-
magnification image ever drawn of the compound eye.

eyes are windows on the soul. How can we fail to shudder at an eye
that cannot weep, turn, or even blink?

There is no reason why natural selection might not, in some corner
of the universe, have thrown up an apt-like monster with large,
bulbous, faceted eyes. Stranger sports have stalked our own planet.
If a snow-elk can get away with antlers far too heavy for its head,
why should there not be an apt-like creature somewhere, looking
out at the world through equally bulky, bulbous eyes?

It is not too hard to imagine a world where every sighted animal,
however big or small, sees through compound eyes. Indeed, for about
85 million years, directly after the Cambrian explosion, compound eyes
enjoyed a monopoly of image-forming vision. Some of them were quite
big: the eyes of the larger sorts of trilobite grew to widths of several
centimetres.

The trilobites were an extraordinarily varied class of animal, ranging from microscopic to the size of a dog. There were stalk-eyed trilobites, trilobites with bulbous visual arrays that dominated the head, eyes so large they met in the middle, eyes hidden away in streamlined recesses; even squitty single facets, the remains of eyes that were degenerating, abandoned as the animal adopted a lifestyle – under the sediment or floating, plankton-like, in deep ocean – where vision was no longer an asset. In 2003, a Moroccan fossil dealer sold trilobite expert Richard Fortey, who works for the Natural History Museum in London, a fossil species new to science which Fortey christened *Erbenochile*. Its eyes look like the speaker stacks of a particularly bad-taste hi-fi system. They came with an integral eye-shade, so *Erbenochile* must have operated in the shallows during the day. Like the rest of its kind, it could not look up, and it did not have to: there were no free-swimming predators willing or able to swoop upon it.

It is fortunate for our understanding of how the compound eye evolved that larval trilobites have left fossil traces. Just as foetal infants, as they develop, roughly recapitulate the evolutionary history of their species, so larval trilobites reveal something about their evolution. The youngest trilobite larvae have widely scattered eyelets pointing this way and that. These rough arrays of light detectors hardly deserve the name 'eyes'. Over successive moultings, these arrays compact and re-order themselves into magnificent, regular compound eyes.

Trilobite eyelets were topped with lenses made of crystal. This extraordinary design of eye vanished with the last trilobite, and the crystal eye has never been reinvented. The mineral employed in the trilobite eye was calcite – the same mineral that makes chalk, lime-stone and marble. In its purest state, calcite is transparent. When lots of tiny microscopic surfaces are compacted – the way fossil fragments are compounded in chalk – light is refracted and reflected to such a degree that the material appears bright white. When calcite grows slowly, layer by layer, as a crystal, it is clearer than glass. There are several forms a calcite crystal can take; a magnificent example of one, Iceland spar, stands in a case on the first floor of the Natural History Museum in London. Look through the sides of this crystal and you will see double: the crystal divides light into two beams.

Another property of calcite is the way it admits light through its length. Light striking the top of the crystal at an angle is refracted to the sides. Only light which shines straight in is carried down its length. This was the secret of trilobite vision.

The chief mechanical challenge for an eye is to gather light from different parts of the scene to form an image. A camera-type eye, like our own, has an easy evolutionary path. Once the eye has formed a pit, closing the walls will produce a pin-hole which casts inverted images on the back wall of the eye. Filling the eye with clear material that focuses the light will improve the image. A lens arising from that clear material will, by moving or deforming, add flexibility to the eye's optical performance. By now the eye works very well and is (for reasons we will come to later) quite large. Having more than a few such eyes would be wasteful and greedy. Humans have two. Some fish have four. Spiders, being greedy devils, have between six and eight, but only two are image-forming.

A compound eye, as it evolves, has no such intuitive path. Consider the choice a primitive array of eyespots has to make at the beginning of its evolutionary journey. So as not to see the same point in space multiple times, the array must point its eyespots in different directions, so that they can report on the light level in different regions of space. A flat plate of eyespots has to deform, and there are two ways it can do this: it can form a dimple, or a pimple. Human eyes are the descendents of dimples. Compound eyes are the descendants of pimples. To begin with, both solutions seem equally good. Only later do pimple eyes reveal their inadequacy and by then it is far too late. Evolution cannot take back its mistakes.

A pimple can never acquire a single, simple focusing mechanism. Eyespots covering a pimple can form an image only by increasing their numbers and arranging themselves ever more neatly and ever more tightly together. Each eyespot, observing a point in space adjacent to the point observed by its neighbour, provides the nascent compound eye with one piece of information (a 'pixel'), and the eye sees a picture made up of as many pixels as there are eyespots. It is an ingenious form of vision, but it has one problem. With no device to focus the light, how can the eye order the image it is trying to capture?

The eyespots themselves, to narrow their field of view, evolve into exquisite and complex structures called *ommatidia*. Architecturally, each ommatidium is a true eye – and a true camera-type eye, at that. In engineering terms, this is a disastrously wasteful replication of effort. (One is reminded of those commercial high-rise buildings in Bombay where the communal air conditioning has broken down and the corridors are crammed with the ducts from countless private units – one for each room.)

An ommatidium consists of a tube, often shielded with pigment and filled with half a dozen or so elongated photosensitive cells packed in next to each other and pressed together at the centre. This photosensitive core is called a rhabdom. The tube is topped by a clear, curved protective coating, which acts as a lens to gather light from just the right part of space, so that each ommatidium has a field of view which abuts its neighbours' without overlapping. The trilobite made a happy choice in calcite – a crystal which passes through its length only that light which hits it head-on.

Although the calcite trick has never been repeated, subsequent innovations have addressed the problem of how to keep each ommatidium focused on one narrow point in space. Pigment shields, graded lenses, light-guides, photoreceptors: the paraphernalia of compound vision still set taxing problems for vision researchers. The pity of it is, of course, that no matter how exquisite an individual ommatidium gets, its allotted task is a dull one: to capture just a single pixel of the view.

Growing tens of thousands of little exquisitely fashioned eyes to acquire a rough, pixellated view of the world seems a ridiculously long-winded way of acquiring vision. Dan-Eric Nilsson once remarked of compound eyes that 'it is only a small exaggeration to say that evolution seems to be fighting a desperate battle to improve a basically disastrous design.'[1] To improve the visual resolution of the compound eye, two things have to happen: first, there need to be more ommatidia, generating more pixels. Second, the pixels have to get smaller: that is, each ommatidium will have to capture an ever-narrower field of view. And this is surprisingly difficult to do, on two counts.

Let's take the easier one first. A lens held at arm's length gathers light from a small patch of space: it has a narrow field of view. Move the lens closer to the eye, and its field of view widens. By the same logic, you can narrow an eye's field of view by increasing the space between its lens and its photoreceptors.

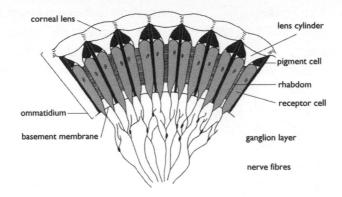

corneal lens
lens cylinder
pigment cell
rhabdom
receptor cell
ommatidium
basement membrane
ganglion layer
nerve fibres

'A basically disastrous design': the apposition-type compound eye.

Human photoreceptors receive light focused through a lens that lies many millimeters away, which means that each receptor is exposed to light entering from a very narrow field of view; about 0.007 degrees. A honeybee's photoreceptor, on the other hand, receives light from a lens that's barely 100 microns away, which means it is exposed to light from a much wider field of view – about one degree. That's an area about the size of your thumbnail at the end of an extended arm. A view made of pixels that size would, as the optical instrument maker Henry Mallock observed in 1894, 'give a picture about as good as if executed in rather coarse wool-work and viewed at a distance of a foot.' The honeybee's eyes are among the best image-forming compound eyes we know of, and the chief reason for their (relative) excellence is the distance there is between an ommatidium's lens and its photoreceptors. As a consequence, the bee's eyes are so bulbous they dominate its head. Were they any larger, it would fall out of the sky.

It is instructive to imagine how big a compound eye would have to be to see as well as a human eye. First, we would have to lengthen each ommatidium by several feet. A delightful and much-reproduced cartoon by the German vision scientist Kuno Kirschfeld substitutes our eyes with compound eyes of equal resolution. It reveals why Burroughs's fearsome apt, if it exists anywhere in the universe, is as purblind as Mr Magoo. If it saw as well as we do, it would have eyes wider than its body is tall and would never be able to raise its fearsome head from the cave floor. Since huge compound eyes are (to say the least)

1 m

Kuno Kirschfeld's fantastical compound eye boasts human-like acuity, but at a ludicrous physical cost.

impractical, evolutionary pressure has tended to favour miniturisation. (Our apt would probably have squitty little eyes, so small as to be barely noticeable.) And miniaturisation raises the second, more complicated difficulty, of narrowing the ommatidium's field of view.

Given two equally long ommatidia of different widths, the narrower will have the narrower field of view. But there are severe limits to just how slim an ommatidium can be. Each ommatidium contains, not just one, but half a dozen or more photoreceptors, which cannot be shed willy-nilly because each one is tuned to a different wavelength of light. This is the basis for colour vision, and colour vision is especially important to small invertebrates, many of whom (the honeybee included) see more colours than we do. The receptors themselves can become narrower, it is true, and the receptors in modern compound eyes are squeezed to their limit: they are virtually no wider than the light waves they are supposed to detect. Many insects tend to see shorter wavelengths of light than vertebrates do, probably so that they can keep their receptors as narrow as possible. But this still leaves the compound eye looking ungainly and bulbous.

The complications set in when we turn to the very narrowest of natural ommatidia – ommatidia so narrow that they run into difficulties with the nature of light.

Punch a small hole through a sheet of paper and shine a light through it on to another sheet. You will see a bright disc where the light shines through the hole, surrounded by a small, faint corona.

Just as sea waves curve around the walls of a harbour, light waves curve around the edges of an aperture. Repeat the experiment, but make the hole even smaller; the bright circle is smaller, obviously, but the corona is correspondingly *larger*. The smaller the aperture, the greater the interference from waves curving around the aperture's edges. This is the phenomenon of diffraction. The smaller a lens gets – and the lenses in compound eyes are just a few times larger than the wavelength of light itself – the fuzzier the image becomes.

It hardly matters whether light entering a single ommatidium is blurred. The ommatidium is only a light detector. The trouble is, when you're as small as an ommatidium, the blurring can be so bad that the photoreceptors find it difficult to detect any light. Light waves can be blurred so much that light captured by one ommatidium can travel through the barrel of another, only to be detected by the photoreceptors of a third.

Some species of insect get around this problem by having pigmented shields around every ommatidium. These don't have to be permanent; at night, some insects can withdraw these shields to improve their night vision.

Other species have hit upon more elaborate solutions. Some do without lenses altogether, opting for elongated 'lens cylinders' with subtle ray-bending properties. (The horseshoe crab has such an eye, and for years investigators agonised over its smooth, lensless corneas.) Butterfly eyes use lenses *and* lens-cylinders. Elsewhere, we find eyes whose ommatidia harness refraction to trap their light. Refraction is the name given to the way light bends when it encounters a boundary between media of different densities; it is why a pencil, dipped into a glass of water, appears to bend. Two factors determine how much the light alters course when it passes from one medium to another. The first is the angle at which it hits the boundary. If the angle is very shallow, the light is not refracted at all, but reflected. This is why panes of glass 'catch the light', making them visible even though they are transparent. The second determining factor is the difference in the densities encountered at the boundary. In water, glass does not catch the light – indeed, it becomes virtually invisible – because water is much closer to the density of glass and thus the refraction effect is much weaker.

The 'pram bug' *Phronima* has found a remarkable use for refraction.

A tiny, deep-sea crustacean, *Phronima* has adopted transparency as its camouflage, but it still has the problem of how to disguise its eyes. Its solution: to distort them beyond all recognition. *Phronima's* eyes are half a centimetre long – a collosal investment for an animal less than two centimetres from from top to tail. Rays of light entering its ommatidia pass through transparent fibres so narrow that the light cannot hit the sides at a sharp enough angle to escape. This effect, called 'total internal reflection', will be familiar to anyone who's ever owned one of those shaggy fibre-optic lamps. *Phronima* has fibre-optic eyes.

Why have more compound eyes not adopted fibre-optics? It sounds like an ideal solution to the miniaturisation problem. Most ommatidia do have modest light-guiding properties, but there is a problem: diffraction imposes an iron limit on how narrow each 'fibre' can be. Light waves travelling along an optic fibre set up standing waves, creating points of utter stillness and points of maximum vibration. The narrower the receptor gets, the more 'dead-zones' it has, until eventually most of the energy of the light wave exists outside the receptor. The narrower the prison becomes, the weaker it gets, until light escapes its bonds completely.

The compound eye is an outlandish and ungainly object; not so much sculpted by evolution as wrenched and twisted. But if compound eyes are so poor, why do most animals still plump for them?

Animals with compound eyes are generally small. The smaller you are, the stronger you are, relative to your weight. Compound eyes may look ungainly, but a small animal is not going to be much inconvenienced by having big eyes (and there is one obvious advantage of having two colossal bulbous eyes – they let you see in every possible direction, with virtually no blind spots).

Being small has other consequences. If you are very small, you are unlikely ever to need to see very clearly, because small things that are far off are probably of no interest. An approaching predator is likely to be much bigger than you, and easily spotted from a distance, even with quite blurry vision. Prey, on the other hand, is likely to be very close by. When a praying mantis is hungry, it simply reaches out its arms for its next meal. Anything of interest to an insect is going to make such a large image in its eye – whether because of its nearness or its sheer size – that high-resolution vision is just a waste of effort.

Humans, being large foragers, tend to fetishise the importance of being able to spot small things. Most other animals are less bothered about images, and concern themselves more with tracking movement. This is another reason why compound vision is a popular evolutionary choice: even the crudest compound eye handles movement well. The very fineness of human eyesight means that moving images blur rather badly, whereas insects, with their crude, pixellated vision, are extremely resistant to motion blur. When your visual world consists of a series of fist-sized pixels, as does the wasp's, there's very little detail to lose. Wasps are among the very few animals whose eyes can see as clearly when they're moving as when they're still (that is, not very clearly at all). They have evolved eyes that are coarse-grained all over, so that they can reconnoitre food sources at speed.

Wasps are not the only species whose vision seems radar-like in its workings. Because compound eyes are not very acute, predatory insects tend not to try and spot their prey against masses of complex vegetation. Instead, they get below their targets, to silhouette them against the sky. Many insects see the sky above them with astonishing acuity. The eye of the dragonfly *Anax junius* – with over 28,500 ommatidia, the densest compound eye of all – boasts a narrow band of very acute vision. Far from caring about images however, *Anax junius* watches for movement through eyes that scan the sky like a radar.

To see what compound eyes are really capable of, we need look no further than the aerial acrobatics of airborne insects. All the compound eye's small but telling strengths come together in flight. Honeybees, as we've seen, have blurry vision, yet they fly with remarkable *élan*, pulling off tricks that put the best stunt pilots to shame. A honeybee's brain is the size of a sesame seed. How do they manage to fly so well? How do they manage to *land*?

If, when you come into land, you keep the texture of things moving evenly across your field of view, you will automatically reduce speed as you descend closer to the ground. If you keep things moving in straight lines, you will automatically descend at a constant angle. You don't need to know your airspeed or your height. All you have to do is keep your eyes open.

In the Thirties and Forties, GC Grindley (known to everyone as 'C') was one of the brightest stars in the field of animal behaviour. Declared unfit for service at the outbreak of World War Two, Grindley – known for his interest in vision and motion – was recruited by the Flying Personnel Selection Committee to study how pilots land their aircraft. He accompanied an instructor, Peter May, on a training flight. May's description of how to land a plane was admirably down-to-earth: 'You look at the point where you want to land and fly towards it till the ground explodes around it. Then you flatten out.'[2]

Back on the ground, Grindley tried to realise May's description in mathematics, and came up with the equations of 'optic flow'. Today, at the All-Weather Bee-flight Facility at the Australian National University in Canberra, those same equations are supporting studies of how bees fly.

Airborne insects judge distances using optic flow. If the bee can't see any texture, it assumes it is flying in open air, and flies quickly. (This is why bees become disoriented indoors and tend to bounce off painted walls: they cannot distinguish between plain surfaces and open sky.) Once the bee notices texture it slows down, and tries to keep the texture moving steadily across its field of view. The closer the texture, the more quickly it moves across the bee's field of view, and the more slowly the bee flies. If images start to move very fast on one side, the bee veers away. A bee comes into land the same way Peter May did; it descends at an constant angle, then levels out at the moment when the images directly ahead 'explode out' towards the sides.

The pattern of optic flow is echoed in the eyes of most flying insects, whose forward-facing ommatidia are narrower and more compact than the ommatidia that look to the sides. This means their forward vision is most acute, while views to the sides are deliberately coarse-grained, to capture the fast-flowing landscape. By dividing the visual scene into substantial pixels of varying width, flying insects have acquired eyes optimised to detect optic flow; that is, optimised for flight.

*   *   *

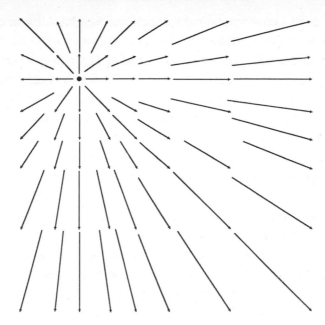

Optic flow: the arrows represent the speed and direction of texure flowing past the photoreceptors of the eye.

Another reason for the ubiquity of compound eyes may be the sheer difficulty involved in trying to change a compound eye to a single-chambered eye without at any stage damaging one's eyesight; the compound eye may simply be too specialised to unpick. Mind you, spiders and scorpions seem to have succeeded, abandoning compound eyes in favour of single-chamber eyes around twenty million years ago. True spiders have eight eyes. Their high-resolution eyes – the best of them as large and as powerful as the eyes of small rodents – point forwards, while the others scout for movements off to the sides. In web-building spiders, the peripheral eyes act as navigation devices. Why have other invertebrates not followed the spider's example?

Some have tried. The first was a mysterious form of trilobite called *Phacops*. Fifty million years after the emergence of the first trilobites, the phacopids made themselves a new sort of eye. With it, they came to dominate the whole trilobite class.

The individual ommatidia of compound eyes capture a single pixel of the view. The more ommatidia there are, the better the eye can see. But *Phacops* rewrote the rule book: it *lost* lenses. However, the hundred or so lenses it retained grew much bigger – some are nearly a millimetre across – and each sat in a small depression, dividing it from its neighbours.

When these eyes were first studied, it was assumed that they were a degenerate form. One typical path for evolution to take is to bring young animals to sexual maturity earlier and earlier. This means that adult bodies can eventually acquire foetal characteristics. In the 1960s it was assumed that Phacopid trilobites had re-adopted their foetal eyes; abandoning compound eyes for scattered eyespots – good for detecting movement, but little else.

But why, if these eyes were degenerate, were the individual lenses so large? At first it was thought that they were large so as to gather light from dim environments. But the figures made no sense. The eyes were actually too good: a single ommatidium boasting a lens like this would be blinded by starlight. It seems much more likely that each lens served a retina made up of many ommatidia. And why were these lenses spaced out so precisely on an hexagonal pattern, with a neatness that put their compound cousins to shame? (A compound eye normally contains a good number of small irregularities.)

In 1972, Kenneth Towe – a palaeobiologist at the Smithsonian Museum in Washington – managed to take a photograph (of the city's FBI headquarters, no less) through the lens of a *Phacops* eye. Towe's photographs showed clear inverted chunks of the building. The images were too clear to be incidental to the eye's main purpose, and made a nonsense of the idea that these were primitive eyes. Working out how the images could be so clear required the assistance of a nuclear physicist, Riccardo Levi-Setti. He discovered that the lenses in Towe's trilobite eye were not pure calcite: a layer of magnesium ran through the middle, correcting their optics and minimising blur. These lenses were more than light collectors. They were *supposed* to capture images.

The trilobite could not possibly have used inverted chunks of the visual scene in their raw state. It must have turned each of these chunks the right way up in the eye and stitched them together. And in that process, it may possibly have acquired another extraordinary optical talent: depth perception. The lenses are arranged so very carefully, and yet their views overlap. Was *Phacops* capable of triangulating the distance of objects from these overlapping views?

If so, this was a truly extraordinary visual system, giving *Phacops* a perception of depth that puts our own, merely binocular, vision to shame. This is too good an idea to leave languishing in the textbooks, and has recently inspired a number of projects in artificial vision. Artificial *Phacops* eyes may yet provide self-steering robots with eyes that can navigate the canyons of Mars, building up detailed maps as they go.[3]

Even supposing that *Phacops* wasn't quite so ambitious (and it is hard to see why it would be), this hardly matters. Whatever uses it had for vision, it saw images, and it had to stitch these images together. It follows, therefore, that *Phacops* would have needed a retina several neural layers deep to process its images. This, in turn, says volumes about what kind of nervous system it must have had.

Phacops was no dummy.

There were other 'experiments', as the compound eye was driven by natural selection to improve its limited design. Chief among them is the 'superposition' compound eye, whose mysterious workings were unpicked by the Austrian physiologist, Sigmund Exner.

Exner was born in 1846, to a wealthy and socially prominent Viennese family. He was extraordinarily fortunate in his education; his tutors, Ernst Brücke and Hermann von Helmholtz, were the two greatest names in nineteenth-century physiology. From 1871, as an assistant in Professor Brücke's renowned physiological laboratory at the University of Vienna, Exner was the mentor of the young Sigmund Freud. Both speculated on the nature of the connection between the mind and the nervous system, and Exner's book on the subject provided the framework for Freud's early thinking about the brain.

Exner's fascination with natural history encompassed the nuts and bolts, as well as the big, imponderable questions. A keen hiker, Exner once observed a buzzard riding a thermal. To understand how the bird managed to rise in the air without moving its wings, he built an experimental buzzard from feathers, clockwork and cardboard. And he once delivered a delightfully oddball lecture which attempted to reconcile aerodynamics with the artistic presentation of flying and floating human figures; angels, *putti* and the like. With a show of perfect sincerity, he stated that *putti* – the winged heads gracing the corners of many a Renaissance painted ceiling – weigh about two grammes and travel at a speed of around two metres a second.[4]

The marriage of engineering *nous* and a nutty imagination served Exner well throughout his career. Who else, at a time when sound recordings were scratched in wax, would have created the world's first sound archive (the *Phonogrammarchiv* in Vienna), laying down recordings which are still playable today? Who else could have written a study of compound eyes as insightful as his 1891 monograph; a text so clear and authoritative that it remains a standard, last reprinted in 1989? This work is more than a text-book: in it, Exner solved one of the most intractable problems in vision – a class of compound eye which seems to refute the laws of optics.

Some compound eyes do not produce multiple images. Their lenses, pressed promiscuously together to form a dome over an eye that is virtually hollow, produce, upon a uniform retina not so dissimilar to our own, a single, *upright*, image.

Fireflies, beetles, moths, and some some species of plankton have

Two forms of compound vision: the individual eyelets of an apposition eye (*left*) each view a different region of space. Each facet of a superposition eye (*right*) bends light selectively onto different photoreceptors. Because it harnesses more of the available light, a superposition eye is better at seeing at dusk and dawn.

turned their compound eyes into single-chambered eyes, called 'super-position eyes', because the images from each lens overlap, producing a single coherent image. It took Sigmund Exner most of the 1880s to work out the horrendous mathematics of such a composite upright image. In doing so, he made a remarkable discovery. The lenses were not behaving like lenses: rather than focusing beams of light, they were *redirecting* them. The individual lenses bent light rays so that they met at a single 'focus' point where one coherent upright image was formed.

How could lenses behave this way? Two convex lenses, carefully arranged, can invert and re-invert an image, effectively redirecting the light; but there were no such arrangements in the eyes of beetles or moths. Even more puzzlingly, the surfaces of the lenses just didn't seem curved enough to converge light rays in the normal way – never mind contort them in ways unknown to optics. Exner realised that variations in refractive power inside the lenses must be performing the inversion and re-inversion. The lenses were not simple affairs but 'lens cylinders' with complex layers.

With no means of testing his idea, Exner set about the calculations that would explain the behaviour of these lens cylinders. A hundred years passed before equipment was invented that could test his conjectures, but these experiments, conducted in 1979 and 1989, confirmed his figures to the smallest detail.

Many superposition eyes see just as well as apposition eyes; the best see as well as a bee. And, with up to three hundred optical elements contributing to the image, the superposition eye is hugely sensitive – up to a thousand times as sensitive as an apposition eye – and ideal for seeing in the dark. It is so good, at least one species, the nocturnal hawkmoth *Deilephila elpenor*, can distinguish colours in starlight.[5]

New forms of compound vision continue to emerge. In 1976, Kuno Kirschfeld discovered that flies and a handful of crustaceans use overlapping views and some complex retinal knitting to sample each point in space seven times. This seven-fold increase in light-detecting ability is great for dusk and dawn vision. Twelve years later, Dan-Eric Nilsson discovered a swimming crab with a unique form of compound vision, involving lenses, waveguides, and parabolic

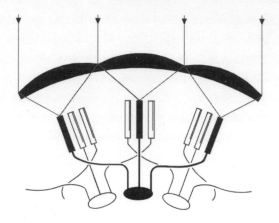

How flies see the world: in a form of vision dubbed neural
superposition, photoreceptors in different ommatidia pool
their signals at the level of the lamina (the insect retina)
to improve vision in poor light.

mirrors. Michael Land of Sussex University has the distinction of
having worked out the optics of the lobster eye, which uses cuboid
mirrors in place of lenses. (A distinction he shares with Klaus Vogt
of Freiburg University – unbeknownst to each other, both had been
working on the same problem.) There will be other discoveries yet.

2 – Leather, water and jelly

In 2003, Anna Gislén, of Lund University in Sweden, grabbed one
of vision research's more exotic projects: she went snorkelling around
Thailand's biggest and most fertile shallow coral reef.

In tow she had eighteen Western children – volunteers from the
40,000 holidaymakers who come to the Mu Ko Surin National Park
each year – and seventeen children from the Moken, a local nomad
people. Gislén was testing the truth of a remark by a senior colleague:
that the Moken can see clearly underwater.

Human vision is lousy in water, because most of the focusing
power of the human eye is achieved by its curved cornea: light passing
from air into an eyeball is refracted strongly. Light passing from

water into the eyeball, on the other hand, is hardly refracted at all, and the human lens is not nearly strong enough to make up for the shortfall in focusing power.

But someone forgot to tell the Moken, who ply the Burmese archipelago and Thailand's western coast. While the menfolk spear fish, the children spend their days diving for clams and sea cucumbers: 'When I first saw them, I was instantly struck by their familiarity with water,' Gislén told the Australian television company, ABC. 'Though it may sound like a cliché, they looked a bit like a school of little fish swimming around.' Simple eye tests – gauging the children's sharpness of vision by asking them to distinguish different patterns underwater – quickly gave experimental weight to the folk tale: the Moken children could see fine detail under water more than twice as well as the Europeans.

The Moken were not superhuman – but they were using the eye to its limits. While the pupils of the eyes of the European children expanded underwater, in response to the dimness of the light, the pupils of Moken eyes shrank to their smallest possible diameter – smaller than European pupils ever go. Though this considerably reduced the amount of light available, it improved acuity no end. Moken children also used their lenses more than European children, squishing them to the limit of human performance.

Gislén is cautious about the implications of her research. 'This behaviour,' says her report, 'is clearly an adaptive strategy.' But after centuries of this kind of life, the Moken people's unique visual ability may just possibly have found its way into their genome. Survival pressures are, after all, the reason why genetic variations are selected and preserved, why communities change and vary, and why, ultimately, new species arise.

The arrival of eyed vertebrates, some eighty-five million years after the Cambrian explosion, introduced an optical innovation so profound, it may be considered a wholesale reinvention of the business of vision. Although mechanically similar to the simple, single-chambered eyes of molluscs and crabs, the vertebrate eye harnessed the available materials in entirely new ways. Its lenses were made out of crystallins, incredibly stable proteins that provided flexibility,

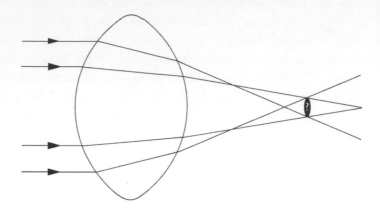

Spherical aberration: the periphery of a lens focuses light more
strongly than its centre, making a blurred image (shaded oval).

transparency and excellent ultra-violet protection for eyes that could
grow bigger, and see the world better, than ever before.

The larger you get, the less power you have, relative to your weight.
Ants can carry five times their own body weight in their teeth; humans
most certainly cannot. The larger you are, the more likely it is, then,
that you feed on animals smaller than you. By the same token, your
predators are likely to be a lot larger than you, and correspondingly
faster, so it is important to spot them from further away. A compound
eye is useless at fine detail and long-range imaging. A vertebrate eye
is excellent at both. Had it not been for the vertebrate eye, living things
would probably have remained lobster-sized for ever.

Even the smallest vertebrate eye boasts a lens that is far, far larger
than any single lens in a compound eye. Its sheer size resolves the
diffraction problem straight away, because the area of focused light
is so large that its accompanying diffraction rings are virtually unde-
tectable. But large lenses have two other weaknesses: spherical aberration and chromatic aberration.

In spherical aberration, the light hitting the edge of a lens is
refracted so strongly that it comes into focus in front of the rest of
the image. One solution is to make the outside of the lens of weaker
optical stuff than the inside. The physicist Heinrich Matthiessen

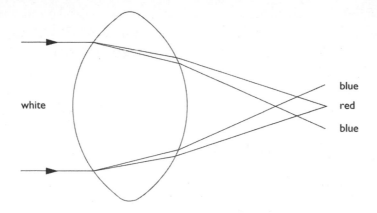

Chromatic aberration: a lens focuses short wavelength
(blue) light more strongly than long-wavelength (red) light.
The figure reflects roughly how the human eye handles
chromatic aberration: red light is brought to focus at the
fovea, while blue light blurs across a wider area.

(1830–1906) worked out the optics of the 'graded index' lenses of
fish, and thereby laid the groundwork for Exner's discovery of lens
cylinders and other optical exotica.

The first vertebrates to venture on to dry land would necessarily
have been extremely short-sighted. Fish, having gone to all the trouble
of evolving powerful, spherical lenses, were now entering an environment where a lens was more trouble than it was worth. Light passing
from the air into such an eyeball is so effectively refracted that the lens
overcompensates, bringing the image into focus much too far in front
of the retina. This is why the lenses of land-based eyes have lost much
of their curvature. It is one of nature's happy accidents that weakening
the lens by drastically slimming it down is one solution to the problem
of spherical aberration. Nowhere, except at the very edges, can light
hit the lens at an angle acute enough to blur the image. This is just as
well because – as we've already seen with our own eyes – the peripheral
image through a slim-line lens isn't that great to start with.

The other major optical problem of large lenses is chromatic aberration. To put it simply, lenses bring the blues of a scene into focus in
front of the reds. The kind of distracting mess this generates can be

best appreciated by looking through a pair of toy binoculars: its cheap lenses, cast out of an homogenous plastic, cannot correct for chromatic aberration, and vividly coloured haloes appear around objects, disrupting their outlines and confusing the perception of depth. The effect is pretty, but you wouldn't want to have to live with it.

Early telescope makers soon learned how to bring all wavelengths into focus at the same point; using a combination of different materials, they were able to produce achromatic lenses. Oddly, although the lenses of eyes are laid down bit by bit, which resolves spherical aberration, no vertebrate lens has ever really addressed the problem of chromatic aberration. Some vertebrates have resolved the problem by adapting the retina: the squirrel's retina is corrugated, so that different parts capture differently coloured focused views. Most other vertebrates don't go to this much trouble. They have smooth retinas, in which the cones are arranged in concentric fields to capture light of different wavelengths.

The human retina has evolved the fovea: a small flattened disc, the very centre of which is crammed with just two types of our three colour-detecting cone cells. Greenish and yellowish light thus comes into focus at the centre of the fovea with nice precision. Our third, blue-detecting cone type is not found the fovea but is scattered among the rods filling the rest of the retina. This is for the very good reason that blue light comes into focus a good half-millimetre in front of the retina, eventually reaching it in a weak, blurred wash. Thus, blue light is always out of focus in the human eye, which may explain why, when we look at a painting, the blues and greens seem to recede, while the yellows and reds leap out. It also rather begs the question of how blue objects in our field of view have sharply defined edges at all? The degree to which the retina has to reprocess the mess of different lights inside the eye, accentuating lines, boundaries and forms, is the subject of Chapter Seven.

Some of the largest eyes in nature belonged to ichthyosaurs. These marine contemporaries of the dinosaurs were themselves descended from land animals, making them the Jurassic equivalent of dolphins. Like dolphins, ichthyosaurs came to the surface to breathe and gave birth to live young. The chances are that their table manners were very like a modern dolphin's: their long thin jaws would have been excellent at capturing fast, agile prey such as large fish and squid. Some early ichthyosaurs had teeth adapted for crushing shellfish.

Of the eighty known species of ichthyosaur, the largest was *Temnodontosaurus platyodon*; more than 15 metres long, *Temnodontosaurus* resembled a great white shark more than a dolphin and, not surprisingly, its eye – at about thirty centimetres across – is the largest known vertebrate eye.

A more familiar ichthyosaur is *Ophthalmosaurus*, named, with good reason, for its eyes. Although smaller in absolute terms than the eyes of *Temnodontosaurus* they were nevertheless the largest in relation to body size. We know a great deal about how *Ophthalmosaurus* saw because of a characteristic of vertebrate eyes not often mentioned: they have bones.

If you are of a certain age you may, when you look the in the mirror, spot an off-white ring around your cornea. If you can, you are looking at pure cholesterol. Where humans lay down cholesterol – within the sclera and around the cornea – most vertebrates lay down bone. For some reason, mammals and crocodiles have lost their sclerotic bones, but every other vertebrate has them, in some form or other. The most complete sclerotic rings we know of – hefty, doughnut-shaped bones, like a fossilised pineapple slice – belonged to the ichthyosaurs.

When a fish-shaped object moves, the front part is pushed by the surrounding water while the flanks are pulled. An eye that's relatively small in relation to body size can be positioned where the forces are about equal. Ichthyosaurs, with their huge eyes, didn't have this option. They reduced the stress on their eyes as much as possible by evolving particularly flat corneas, so that their eyes were relatively flush with their skin, but this made their eyeballs an odd shape.

So, the ichthyosaur's shallow eyeballs needed scaffolding. This is very good news to paleontologists, since fossilised sclerotic rings provide them with reliable measurements, not only revealing how old an ichthyosaur specimen was when it died, but also how much light its eyes could gather. *Ophthalmosaurus* is well-named: it could see in the dark – and were it a land animal, the obvious conclusion would be that it was nocturnal. For an air-breathing aquatic animal, this is very unlikely; *Ophthalmosaurus* would have to have slept on the ocean surface in broad daylight. It is much more likely that it was a deep diver, hunting for prey in the dark waters more than half a kilometre below the surface.

This is still not enough to explain why it had such large eyes. Modern seals have relatively small eyes, and they hunt just as deeply for their food. No one has yet explained with certainty why ichthyosaurs boasted such large eyes. Perhaps it says more about their prey than it does about ichthyosaurs themselves: the smaller and more manoevrable their prey, the larger their eyes would need to be, to see and track a meal at such depth.

The vertebrate eye evolved in the ocean, and when it moved on to dry land, it took a little of the ocean with it. It had no choice: the tissues from which the vertebrate eye is made need to stay wet to survive. Even while it was still in the water, the vertebrate eye had acquired a number of protective devices, any and all of which were capable of protecting and cleaning the eye once it left the seas.

These were not 'eyelids' as such; the water-dwelling eye has no need for the wiper-blade of a moving eyelid. Rather, they were spectacles, examples of which can be found throughout the natural world, on land as well as in the seas.

The corneas of lampreys exude transparent membranes: ideally, these lie against the cornea proper, making them quite hard to spot. If they are damaged, they can peel away, with no damage to the cornea beneath. Bottom-feeding fish have their eyes permanently shut, hidden beneath a layer of translucent skin that protects them against sand and debris. Some fish with bulging eyes wear translucent skins over their eyes, to improve their streamlining. Snakes and other reptiles are descended from animals that had eyelids more or less like ours. Confronted with constantly blowing sand, however, their eyelids have fused over and become transparent.

For most land animals, cleaning the eye is as important as protecting it; hence moving, semi-rigid eyelids. Look into the near corner of your eye: the little pinkish fold there is the last remaining nubbin of your third eyelid. Many terrestrial vertebrates have hung on to these inner eyelids – called nictating membranes – and their uses are legion. Beavers and manatees use them as a barrier to reduce inflammation when they go swimming; seals peel them back while they swim, but slide them over their eyes on land to clean off sand and other debris; the aardvark uses them to shield its eyes from its termite prey; birds of prey use them to protect their eyes from their chicks while they are feeding them, and in polar bears, the nicitating membranes protect the eyes from snow blindness.

So much for our wet, delicate and almost unbearably vulnerable eyes: looked at from an evolutionary perspective, the vertebrate eye is one of life's most resilient adaptations, capable of resolving the world with exquisite clarity at depth, in desert storms and Antarctic white-outs, under the brightest of bright lights, and in near-blackout.

Retinas are sculpted by the kind of light they are exposed to. Blurred light encourages visual abilities that do not rely on clear images: areas devoted to the detection of movement, and to providing good all-round vision in dim light. Well-focused light favours visual abilities that make use of plentiful light and a fine-grained view of the world. This is why the photoreceptors of the vertebrate retina have evolved and specialised into rods and cones. Rods can be triggered by light coming from all directions; their resolution is poor, but they are very good at gathering light. Cones are different; like little optical fibres, cones gather only the light that hits them straight on, and cannot be so easily triggered by stray light. Only well-focused light, entering the eye through the centre of the pupil, will stimulate them. This makes them useless in the dark, but very good at seeing the world in detail during the day.

For rabbits, the whole horizon is in reasonably good focus, and they become aware of potential predators the moment they appear. They barely have to move their eyes to get a bead on potential threats, and their eyes perform very few saccades. They do not have foveas, because they do not need them. Neither do cats, even though their eyes face forwards and they have to judge distances with some accuracy. Cats pay more attention to the horizon than to a specific focal point. Like the retinas of rabbits, cat retinas boast 'visual streaks': lines of enhanced acuity, aligned to the horizon.

Seabirds are among the many animals that have both a visual streak and an area of good central vision (not nearly as specialised as a primate fovea) called the *area centralis*. Birds of prey have a fully-fledged fovea; some birds have two, which enables them to watch what they're eating at the same time as they're watching the world.

The vertebrate retina has adapted over time to serve some truly bizarre visual requirements. My favourite is the surface-feeding fish *Anableps anableps*, a four-eyed fish from Central and South America which greatly entertains my daughter when we go to our local aquarium. It scuds along the surface looking for flies: thanks to its double eyes, it can see clearly in and out of the water.

Human eyes are no less specialised, and no less odd. Only animals who forage in a three-dimensional environment, for example, the forest canopy, can afford a retina that is not in some way tuned to the horizon. In humans, who have returned to the plain, this horizonless style of vision, far from falling away, has been taken to a pitch of extraordinary specialisation. Only a handful of birds of prey have foveas with finer acuity than our own – and they have to detect prey hundreds of metres away on the wing. Why are humans, of all land-dwelling vertebrates, virtually eagle-eyed?

This account from 1896 holds the clue:

An expert tool juggler in one of the great English needle factories, in a recent test of skill, performed one of the most delicate mechanical feats imaginable. He took a common sewing needle of medium size (length 1 5/8 inches) and drilled a hole through its entire length from eye to point – the opening being just large enough to admit the passage of a very fine hair. Another workman in a watch-factory of the United States drilled a hole through a hair of his beard and ran a fiber of silk through it.[6]

Without hawk-eyed vision, we would be unable to perform close work. We would not be able to make tools. We would not be able to arrange and organise our environment. We would not be able to make fine distinctions between objects, spaces, species, peoples, and individuals. The very richness of our social world, full of manipulatable things, rather than fleeting, corner-of-the-eye impressions, has driven the human eye towards high acuity vision, and continues to do so. Since the 1280s, humans have been grinding lenses and sticking them on wires in front of their eyes. Now we stick the lenses onto our eyes: last weekend I went for a dive wearing soft plastic self-balancing contact lenses, which correct for the slightly uneven shape of my eyes' natural lenses (a common condition, astigmatism). When I came out, I plucked those two tiny, barely visible miracles of materials science off my corneas and tossed them into the ocean: they were only disposables.

We look through telescopes and microscopes. The more foolhardy of us allow our corneas to be sculpted by lasers and spinning blades, on the promise that we might get to see the world in finer and finer detail. There is much evidence that, in our hunger for good tool-making vision, we are steadily blinding our species, and becoming, as a consequence, more and more reliant on technology to prop up our ailing sight. Myopia rates are rising alarmingly. In 1996, about sixty per cent of American twenty-three to thirty-four-year-olds were short-sighted, compared with about twenty per cent of people over sixty-five. In Asia, things are much worse. The Singapore National Eye Centre estimates that more than eighty per cent of the country's eighteen year-old men are myopic.[7]

This matters. In the developed world, degenerative myopia is a leading cause of blindness. People who (like me) wear lenses stronger than six dioptres are more susceptible to glaucoma and more likely to suffer a detached retina. Although susceptibility to myopia can be inherited, the environment is most to blame. The bald fact is that illiterate populations need fewer spectacles. In rural Nepal and the upper Amazon, only one in a hundred is short-sighted. There is even evidence that children grow more short-sighted in term-time than during the summer holidays. Many epidemiologists blame the current myopia explosion on families (families are a favourite soft target for this sort of hell-in-a-handcart punditry): children should be getting out more, and playing on their PlayStations less!

No one ever suggests closing down the schools.

FIVE

Seeing and thinking

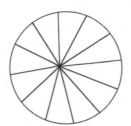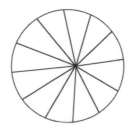

About a year ago, I was sitting in a cafe when an exhausted-looking woman sat down beside me, took her baby daughter out of her pram, and set her down on the chair beside her. The baby – and she really was only a baby – climbed off the chair, dropped neatly to the floor, and started running around the room.

Impressed, I asked her mother how old she was. 'Ten months,' she replied. "REBECCA, MIND THE . . .!'

Rebecca bounced off the leg of a nearby table, giggled, turned, and hurled herself at a wall. Thanks to her prodigious ability to walk – *run* – before she was even a year old, Rebecca demonstrated an important fact about human vision to me: for the first year or so, it is distinctly unreliable.

During its first month of life, a baby's eyes can do little more than protect themselves. An object moving on a collision course with its eyes will make the baby blink. At three months, the baby uses the motion of objects to work out where their edges are; a month later, the baby will use both motion and stereo vision to work out how the edges relate to each other, beginning, for the first time, to see in three dimensions. At seven months, a baby can use shading, perspective, and occlusion (the way some objects obscure others) to perceive depth and shape. Now – and only now – is the child is old enough to recognise familiar objects by sight alone. At twelve months, they begin to name things.

This last point, impressive as it is, may seem irrelevant to the business of vision, for surely there is no obvious or direct connection between vision and language?

The cognitive and computer scientist Donald Hoffman thinks there is. He has studied how we divide the world into parts, and how we give those parts names, uncovering a set of 'rules' by which we divide surfaces and lines into things and parts of things, and he is finding ways to express these rules in numbers. He explains: 'The way we carve up the world verbally is not arbitrary; it depends in part on how we carve it up visually. And the way we carve up the world visually is not arbitrary; it depends in part on fundamental principles of mathematics.'

How should we interpret Hoffman's work? He has come up with a workable model of how we see things. But that does not mean our eyes really work like that. The American psychologist James Jerome Gibson (1904–1979), whose most influential papers were published in the 1960s, pooh-poohed the idea that perception required any kind of calculation. According to Gibson, the mind does not need to assemble the visual world out of single points of light. It does not need means (rules) to achieve ends (vision). All it has to do is 'resonate' to the many salient patterns light contains.

My guess is that somewhere between the maddening vagueness of Gibson's 'ecological school' of vision and the maddening prescriptiveness of the computational school (for which Donald Hoffman is a recent and able exponent) there exists a good description of how seeing is related to thinking. I have neither the space nor the talent to do the job myself. I can do no more than conduct a whistle-stop tour of the mysteries and problems surrounding the seemingly mundane act of 'seeing something', and suggest ways in which seeing, thinking and language are related.

The first part of this chapter, 'Seeing things', looks at what is, even among vertebrates, a rare talent: the ability to divide the world into *this* and *that*. How are the same objects spotted and recognised from very different angles, under many different lights? The second part, 'Seeing and remembering', looks specifically at human vision, which has taken the business of identifying objects to a curious level of specialisation. The final part, 'Reading each other', looks at the eye's role in communication. The several species on Earth that employ language all have excellent spatial vision (they have something to talk about) and are all social (they have someone to talk to). I tell the story of efforts to unpick the relationship between good vision and social living in our own species – and why that relationship may have encouraged the development of language.

1 – Seeing things

Konrad Lorenz (1903–1989) discovered his vocation early. Each evening, when he was a small child, one or other of his parents would read to him from *The Wonderful Adventures of Nils*, a children's book by the Nobel prize-winning novelist Selma Lagerlöf. It tells the story of Nils Holgersson, a boy turned by magic into an elf. Happily for science of animal behaviour, elves ride on the backs of wild geese: 'I yearned to become a wild goose,' Lorenz remembered, on receiving his own Nobel prize. '. . . on realizing that this was impossible, I desperately wanted to have one and, when this also proved impossible, I settled for having domestic ducks.'

From these charming, humble beginnings, and from the 'shattering disillusionment' that accompanied his adult study of the literature of animal behaviour ('None of these people knew animals', he remarked. 'None of them was an expert.') Lorenz constructed a new science: the study of animal behaviour.

A particular interest was the way newly hatched birds 'imprint' on the first big moving object in their visual world. Unlike humans, who seem content to wobble blearily about for their first year, newly-hatched goslings wire up their visual worlds with impressive urgency. Haste, of course, can lead to mistakes, and for the goslings who imprinted on Lorenz, the scientist became 'mum' (and more – the same animals later found him sexually attractive).

In 1936, Lorenz met Nikolaas Tinbergen, the restless younger son of Dutch intellectuals. Tinbergen, with a catalogue of mediocre school reports and lax university attendance behind him, was approaching his thirties, trying to turn his childhood interest in rambling and nature study (anything to get out of the classroom) to some account. His eccentric and original approach to biological problems, and especially his insistence on the value of field study, found a ready audience in Lorenz. Ostensibly, their relationship was what of master and pupil, but Tinbergen's irrepressible urge to check Lorenz's hunches by experiment soon elicited, Tinbergen later remembered, 'an almost childish admiration' in the senior partner.

It is significant, incidentally, that Lorenz and Tinbergen favoured birds for their studies. Birds are easier to understand than mammals because, they, like us, depend largely on their eyes for social contact; most other mammals think through their noses. The biologist Sir Julian Huxley once remarked that if we were olfactory animals, there would be no bird watchers: we would have mammal-smelling societies instead.

Tinbergen set out to find the simplest stimuli that would provoke a response in the birds under Lorenz's care. If he dangled a scrap of

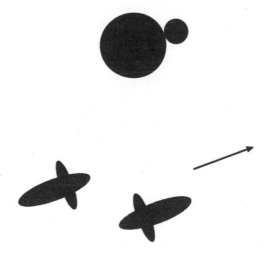

Top: baby blackbirds recognise a mother blackbird
in this arrangement of circles.

Bottom: in this shadow-play suited to the visual proclivities of
ducks and chickens, a hawk (*left*) harries a goose (*right*).

black cloth from his fingers in front of his jackdaws, they reacted as though one of their number was being attacked. Young blackbirds, presented with two discs of different sizes (one about a third of the diameter of the other) treated the arrangement as a parent. Chickens and ducks distinguished hawks and geese as crosses: the goose has a cross-piece at the back; hawks have a cross-piece at the front.

We may smirk, and think that these animals were easily fooled, but if we do, we miss the true glory of Tinbergen's discoveries. The world is not full of flying crosses or carefully proportioned black discs but it *is* stuffed with parents, siblings, predators, mates, and prey. To recognise these things from every possible angle, under every possible light, animals make generalised models. Our experiments fool them only because we've presented them with a model that closely resembles their mental model.

We can fool ourselves in exactly the same way. The Necker cube (below, left) is fun because it 'flips'. The figure has no shading or occlusion, so there are two ways of interpreting how it lies in three-dimensional space. Is that a near edge, or a far one? Is that surface seen from above, or below? The mind cannot decide; it can only juggle the alternatives. This is, quite frankly, weird, not least because both answers are quite spectacularly wrong. The right answer is obvious: the figure lies flat on the paper. We might know that the figure is flat, but we find it extremely difficult to *see* that it is flat. Only when we tweak it (below right) so that its near and far corners are perfectly aligned – a coincidence too great ever to occur with any frequency in the real

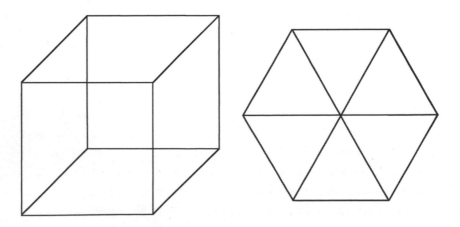

world – do our eyes accept the obvious: the figure is two-dimensional.

On a busy, changeably lit planet, where objects approach us from all angles, set against goodness knows what backgrounds, seeing involves making assumptions. The more limited your surroundings and behaviour, the less you need interrogate the world, and the larger your assumptions can be. Brains are ruinously expensive to run, so the fewer brains an animal requires, the better. A hedgehog can attempt to seduce a toilet brush, and its long-term breeding chances are not much affected. For a human being, a mistake on this scale would be catastrophic. We are a social animal, and we expect more from our mates.

We are still capable of error. To see is to surmise, and error is always possible. Konrad Lorenz recalled one such mistake during his Nobel lecture: 'Once I mistook a mill for a sternwheel steamer. A vessel was anchored on the banks of the Danube near Budapest. It had a little smoking funnel and at its stern an enormous slowly-turning paddle-wheel.'[1] This was no different in kind to the mistakes made by Tinbergen's baby blackbirds. Seeing things involves making assumptions and that requires having an idea of what they might be.

It may not be much of an idea. The pathway from retinal pattern to idea and back to visual impression is such a well-trodden one that we're no more aware of it than we are of the beating of our hearts. As Hermann von Helmholtz wrote: 'The psychic activities that lead us to infer that there in front of us at a certain place there is a certain object of a certain character, are generally not conscious activities, but unconscious ones.'[2] An idea it is, none the less – and this raises an interesting and by no means easy question: at what point does seeing become thinking?

2 – Seeing and remembering

Different animals have different fields of view. Some fish have an excellent vertical field of view, and are able to observe the world both above and below. Some birds and insects have eyes positioned so that they can see all the way around; they have a 360° horizontal field of view. The horizontal field of view enjoyed by humans is half that – about 180°. Of that, the central 140° are observed by both eyes.

When I stick my thumb out, at arm's length, its width covers about two degrees of my visual field. My eyes can bring an area slightly less than that into perfect focus.

Just one degree away from the very centre of my vision, I see things only half as well. Five degrees off centre, my 'visual acuity' is quartered. Beyond that five-degree radius, I can no longer be sure of what I am seeing. At twenty degrees from the foveal centre, my visual acuity falls below the common legal standard for blindness.

Our eyes bring little bits of the visual world to our attention, and from these shards we build our world.

Even as I write this, knowing it to be true, I balk at its oddness. I look up from the keyboard and I turn my head, just once. In that glance, I see a world of vivid colour, a world of three dimensions, of solid, textured forms. I see a world of objects: sofas and screens, floorboards, a pair of sunglasses on the edge of my desk; a reading lamp. Many of these objects are made of parts, and I see these parts, and how they are fitted together: the lamp base, the switch, the bulb. Outside my window there is a tree, its every leaf clear and distinct. A car passes on the street beyond the tree, and I know, even from this brief glance, what sort of car it was, what colour it was, and even, roughly, how quickly it was moving.

I get out of my seat. The car is gone, but the lamp is here: I can touch it. The bare floorboards are solid beneath my feet, as I saw they would be. The wooden lattice-work screen that belonged to my wife's mother casts speckles of light on the bare polished floor. It is a bright day, and if I put my hand in front of the screen, the light makes little pools of warmth on my skin. Everything is as I saw it. Everything confirms the astonishing accuracy of my casual glance.

Impossible to imagine that I did not actually see any of this. Impossible to imagine that I made it all up. If I made it all up, how could I have got so much of it right? How can this illusion be so accurate as to withstand the test of exploration? If I really can't tell, from my peripheral vision, that my sunglasses are perched precariously on a corner of my desk, how come, when I turn my head to study them, there they are, right where I thought they were? How come they are, indeed, sunglasses, and not some other, quite unexpected object?

Is there, somewhere in my head, a model of the world; a mental model so accurate it includes my sunglasses? Maybe, without my knowing it, every little degree-wide glimpse of the world contributes to a vast mental model. Maybe that is what my model is – a composite of little glimpses, memorised with perfect fidelity. The early pioneers of robotics thought so, and built their robot eyes and visual systems accordingly. But most of their robots never moved, and most of the ones that did move, fell over. The others walked into walls. Slowly. The computational demands required to hold and move about in a pin-point accurate, three-dimensional model of visual space are not just high (putting them beyond the capabilities of robots) they are outrageously high, far exceeding the capacity of our brains.

Yet mice, with their little mouse brains, do not fall over the furniture; birds do not fly into brick walls.

How do they manage it? How do we manage it?

It is 1997; the British artist, Emma Kay, is writing the Bible.

'The disciples were Matthew, Mark, Luke, John "The Baptist", Peter "The Rock", Paul, Simon, James, and three others.'

She is sitting at a big, white desk. Her manuscript, written in longhand, rests in front of her. It's a slim volume: by the time she is finished, it will only be about 7,000 words long. This is the Bible according to Emma Kay: rather shorter than the original, and not terribly accurate. 'I had a vague recollection that Elijah was fed by ravens,' she explained to Rosie Millard for *The Tastemakers*, a book describing the Brit Art scene in 2002. 'I had absolutely no idea why or how or where it fitted in, but it had to go in somehow.'[3]

The following year, Emma Kay wrote the complete works of Shakespeare; the year after that, she wrote a history of the world, which has since been published under the title *Worldview*. It's 80,000 words long – as long as the average novel – and it's a complete failure. Which is the point. Emma Kay doesn't use reference books. She relies on her own memory. (This is why her work tells us far more about her – her childhood, her upbringing, her interests, her personality – than it does about the Bible, or Shakespeare, or the history of the world.) Emma Kay's work asks the question: how much do we really know about anything? Her work is certainly amusing. It is also humbling. Spawn of cheerful atheists,

I'd never heard of Elijah's ravens. What would my Bible look like?

How do we know what we know? Do we know something because we remember it, down to every last detail? Or do we know it because we can lay hands on it – because we know how to know it?

Why hold all the world in mind, when we can simply flick our gaze from thing to thing? Why remember, in exquisite detail, which objects are which, and where, and in what relation, when our eyes can rake the real scene in moments, giving us, every third of a second, a nugget of fresh information?

Did I know my sunglasses were perched precariously on the edge of my desk? Even as I asked the question, my eyes were interrogating that corner of space. They got there before me. Even as I asked the question, my glasses were in view.

Perhaps the most elegant scientific demonstration of the eye's 'search function' was devised by Dana Ballard. Ballard is one of a new generation of robot builders. Robots have come a long way from their pioneering days. They no longer carry models of space around in their tiny, painfully-crammed 'brains'. Instead, they mimic as closely as possible the behaviours of real animals. Their eyes move, inter-rogating the world for the information they need.

Ballard wanted to know what information his robots needed to see, how the information should be stored, and for how long. On the assump-tion that 538 million years of evolution must be good for something, he began by looking at how human eyes behave when engaged in simple tasks. He asked a group of volunteers to copy a geometrical pattern, using a set of coloured blocks. Looking closely, this is what he saw:

First, his volunteers glanced from the pattern in front of them to the tray of loose blocks. Next, they picked up a block of a certain colour. Then, they glanced at the pattern again. Finally, they fitted the block they had selected into the position required to make a copy of the pattern.

This seems just a bit complicated: if the eyes see a block of a certain colour in a certain position, why can't both pieces of infor-mation be retained in one go? Why does the eye make one set of movements to choose a block of a particular colour, and another, virtually identical set, to establish where it should go?

The answer seems to be that it is easier for the eye to move than it is for the mind to remember. Why fill your head with transitory and trivial information about coloured blocks, when the eye can simply glance at them, again and again? The eyes are our search engines, cheaper and faster than memory.

After a lifetime of pen-pushing for the local electricity board, quite out of the blue, my father acquired an enthusiasm that would dominate the rest of his life. One day, when he was at a loose end, he volunteered to help fetch and carry for a local archaeological team who were excavating the layout of some medieval gardens. A line of Roman foundations and an Anglo-Saxon burial later, he was hooked. And this is how I came to spend the weekends of my early adolescence traipsing across freshly ploughed farmland in the rain.

Field-walking is the least glamorous, most miserable activity in archaeology. At a dead march, head bowed, eyes fixed to the mud and grit beneath your feet, you plod, endlessly, for hours, watching, waiting, for something, anything, to catch your eye. We walked the chalklands of the South Downs, watching for Samian ware pottery, for the cores of worked flints; if we were lucky, a scraper or two.

Flint tools were a highly developed, very precisely worked technology. The raw material was carefully selected and highly prized. Flints and flint tools were traded across Europe. A flint scraper found in East Sussex might have been struck from a core imported from France.

So here am I, in 1977, twelve years old, galumphing blindly over the remains of this rich and subtle past, and I'm picking up anything. Everything. Stones. Bits of brightly-coloured earth. I haven't got a clue. Flints fracture in the cold: morning frosts shave splinters off dark, greasy, glassy flint and scatter them for nerdy twelve-year-olds to cut their frozen fingers on. Dad shrugs. 'Frost damage,' he says, handing the flake back to me. Stupidly, I toss it aside. As we work back across the field, I will spot it again, and make the same mistake. (It's the hope that kills you.)

Then, after a few weeks, something strange happens. Is that the worked core of a flint? *Yes.* Is that a piece of pot rim? *Yes.* They are starting to reveal themselves, emerging from the surrounding chaos. They are making themselves known.

They are catching my eye.

Twenty-five years on, I am sitting in the office of Michael Land, the man Richard Dawkins dubbed 'the King Midas of eye research'. I can hardly write a book about eyesight and not meet the man who discovered some of the more unusual eyes in nature; who unpicked the workings of the lobster eye, notorious for apparently contradicting the laws of optics; who was one of the first to see saccadic patterns in the flight-paths of small airborne insects; and who, with Dan-Eric Nilsson, wrote *Animal Eyes*, the bible of biological vision.

Quite how we have ended up talking about orchids, I'm not sure. In addition to his own work – never mind the help he gives other researchers, keeping up a lively correspondence with everybody who is anybody in the field of vision research – Michael Land finds time to explore English woodlands for rare orchids.

Like the pot-sherds of my childhood, Land's orchids are hard for a novice to spot. It is only over time that they begin to stand out, and draw the eye.

What has happened in Michael Land's head, that he can spot a rare orchid at twenty paces?

Come to think of it, what has happened in your head, that you are able to read this sentence?

When we read, our eyes fixate on between just twenty and seventy per cent of the words. Who or what tells our eyes which words to choose? Where does that knowledge reside? How is it made, where is it stored, and how is it accessed?

Land, who has always found field-work more congenial than lab work, is looking for the answers to these questions in everyday situations. Typical of his hands-on, artisanal approach to his subject, his experimental gear is home-made. 'Where was I going to find thirty thousand quid to buy one of these from the States?' he asks, as he displays – with no little pride – an eye-tracking device of his own invention. This cost £1,000 to make and colleagues from richer climes (Land works in a cupboard-sized office in the University of Sussex) can never understand where he's hidden all the computerised bits. That's because there aren't any. Land uses a tiny mirror to split the images recorded by a small head-mounted videocamera. The movements of the wearer's eyeball fill one half of the video image; their view of the world appears on the other. Land and his colleagues correlate these images by hand.

These cottage-industry records are some of the most revealing visual documents of eye behaviour ever produced, and allowed eye-movement research to leave the laboratory and enter the real world.

The quotidian character of his findings have earned him headlines. It was Land who discovered that, when a ball is bowled at a cricketer, the cricketer takes his eye off the ball and looks instead at where the ball is about to bounce. Similarly, when negotiating a turn, a driver's eye will provide motor information to her arms almost a second before any muscular movement is required. It seems, from Land's studies, that half a second of visual information is held in a 'buffer' of some kind. The eyes stay one step ahead of the body, dealing with the next view, the next task, the next set of predictions and calculations, while the body relies on the 'buffer'. This raises the odd but compelling idea that the 'present moment', as we experience it, has a measurable duration. We operate in the world, not as it is, but as it existed half a second ago.

How can the eye stay half a second ahead of us? How does it know where to go?

Throughout our lives, we learn to see things afresh, adapt to new experiences and spot novel objects; be they pot-sherds, orchids, flying cricket balls, or fast-approaching bends in the road. But it has always been assumed that this 'perceptual learning' is an advanced business, involving many areas of the brain. Land's recordings show that, on the contrary, this 'learning' must be an excessively simple affair, that enables the eye to recognise and respond to salient features of the scene long before they are ever brought (if indeed they are ever brought) to our conscious attention.

When we say that 'something catches our eye', we are being less fanciful than we might imagine. The eye is not our servant. It is our ambassador. In the tug-of-war between Self and World, the eye is the red ribbon at the centre of the rope.

Michael Land fell in love with orchids and wanted to spot them in their natural habitats, and this desire set in train a process of learning. But it is the plants themselves – their colours, their appearance in different lights, their textures, the opacity of their petals – that have taught him to spot an orchid in an English wood.

Our eyes interrogate the world, and through our eyes, the world announces itself.

3 – Reading each other

The eyes of men converse as much as their tongues.
                                                  Ralf Waldo Emerson

Last year, while visiting New York, I made time to visit the FAO Schwartz toy store. I returned home with a box of interconnecting plastic shapes called Zolo. As you fit these shapes together, you find yourself making little creatures.

The shapes aren't anatomical. You don't say, as you tip the pieces out: 'That's an arm, that's a torso – look! an ear.' Yet after a few minutes' play, and almost in spite of yourself, you will find in your hands a cute, crazy little alien thing with an expression and a character all its own.

The reasoning is simple. Among the shapes are a handful of small white balls, each with a discreet black dot. These lifeless things are *eyes* – though they bear precious little resemblance to any natural eye. White balls with black dots: somehow they look like eyes, even before you attach them to other pieces of Zolo. Added to whatever shape you happen to be building, they make the thing – however random – come instantly to life. That curly swizzle-stick becomes a tongue; that red ball, a nose.

Why does this happen? What gives this simple shape – a dot within a circle – such power over our imaginations?

In a world where every animal is looking for something to eat, the first object a sighted animal must learn to recognise is the eye itself. Being looked at matters. If something's going to fight you, or eat you, or try to mate with you, it's going to be looking at you.

Stare at a hog-nosed snake from a distance of about a metre, and it will play dead. Turn away, and it will move. This behaviour – tonic immobility – seems rather futile to us, with our object-centred, stereoscopic gaze. But since most animals watch for movement and care little for stationary objects, the strategy works well in most cases.

Lizards, chickens, blue crabs and ducks all exhibit this behaviour, and its traces are still strong in us – as anyone who's ever played 'grandmother's footsteps' as a child can testify.

Every eye is on the watch for other eyes. Whether it wants to be or not, the eye is never neutral. It does not simply observe: it communicates.

Long before the cave artists of the Ardèche, in southern France, turned their cave walls into galleries of bison, horse and mammoth, the process of natural selection had thrown up the planet's first artworks. The eyes adorning a peacock's tail, the eyes on a hawk moth caterpillar, the eyespots on the tail fin of the four-eyed butterfly fish *Chaetodon capistratus* – these are, in every sense, pictures, works of *trompe l'œil*, there to confound and impress the onlooker.

Few animals risk making a feature of their real eyes. The banded butterflyfish *Chaetodon striatus* goes so far as to hide them within a black bandanna. The 'whites' of most vertebrate eyes are not white at all; many develop an all-over dark pigment that effectively conceals where, precisely, they are directing their gaze. Only an extremely social animal would make its eyes more noticeable than they had to be. Humans are exceptional in having bright 'whites' to their eyes precisely so that an individual can tell where its fellows are looking. In a world where every eye is primed to spot every other eye, this is a risky adaptation. What benefit can it possibly bestow?

Answering this apparently simple question involves a story so convoluted, I can only beg the reader's patience.

To begin at the beginning: 1957 was a bad year to be a rat. In laboratory after laboratory, experimental rats were collapsing in convulsions, and nobody could work out why. The problem attracted the particular attention of the zoologist Michael Chance, one of a team of bright young researchers brought to Birmingham University by Joel Elkes, generally regarded as the father of experimental psychiatry.

Contrary to the clichés of the animal liberation lobby, lab assistants tend to be sentimental creatures and grow very attached to the animals in their care; certainly the rats Chance examined were healthy and well tended. And yet the instant an assistant unwittingly jangled his or her keys – over they went, in spectacular spasms. Something was making otherwise healthy rats behave *as if* they were convulsing. Realising this, Chance very quickly spotted what was wrong. In the wild, a startled rat runs away and hides. Even a pet rat generally has a sleeping box, where it can conceal itself. No one had thought it necessary to give laboratory rats sleeping boxes.

The moment sleeping boxes were installed in the cages, the rats stopped convulsing. When Chance jangled his keys, they did what rats normally do – they ran away and hid.[4]

At the time, little was known about the emotional lives of animals. The dominant psychological doctrine of the time, behaviorism, regarded the 'inner states' of animals and people as having little scientific value. They could only be by-products of reflex behaviours, fortuitously generated phenomena which played no useful role.

But if rats were simple switching systems, swapping behaviours the way a toy robot changes directions in a maze, why were stressed rats such lousy subjects? Chance showed that animals kept in conditions closer to their natural environment produced better experimental results, while animals stressed beyond endurance not only behaved abnormally; they lived abnormally, slept, ate and moved abnormally, and any measure of their condition during an experiment was more likely to be a product of stress than anything to do with the experiment.

For Chance, the study had two consequences. It drew him into pioneering work to improve the living conditions of animals – especially primates – kept for laboratory study, and it led him to think more about fear.

Every animal, faced with a deadly threat, resorts to some sort of random behaviour. Many animals rely almost entirely on unpredictability: rather than head straight for cover, a rabbit pursued by a fox will bob and weave in a chaotic zigzag. Confused animals make poor predators, so unpredictability is a very good defence, cheap to develop and easy to deploy. (When they convulse, rats become difficult to catch and hold. A cat that has not been taught by its mother how to kill its prey will try to hold a mouse in its jaws; the mouse convulses and almost always gets away.)

The puzzle comes when we look at our own unpredictable behaviours. And we have them, in abundance. We lash out. We react in unpredictable ways. We fool, and we cheat, and we lie. The odd thing is, we behave like this all the time. We are the moodiest animal on the planet, deploying, with great finesse and subtlety, every waking moment, behaviours that in other animals are crude, last-ditch, and above all, short-lived bursts of anomalous behaviour.

Except in primate societies, conflicts never persist in nature, and squabbles between animals of the same species rarely become serious enough to trigger truly unpredictable behaviours.

Chance, together with another Joel Elkes protegé, Allan Mead, was the first to realise that humans are different: they regularly use unpredictability. In the light of Chance's study of lab rats, this observation had daunting implications. It meant that primates had acclimatised to, and learned to master, a condition of perpetual terror. This was a kind of evolutionary pressure that no one had ever considered before. Was terror a driver of human evolution?

We tell our children, in school and in our books and science programmes, that human intelligence arose because we had to develop tools to survive, and language arose because we had to communicate and co-operate with each other. No doubt these things are true – but are they enough to explain human intelligence? Many primates use tools, and communicate well. The more we learn about animal behaviour in the wild, the more instances of language and tool use we find. Why did human language and tool use become so different, in kind and scale, from the language and tool use of other animals?

A veteran of forty years of lemur research, Alison Jolly first visited the island of Madagascar in 1963, after studying at Cornell and Yale. Ask her why she chose the lemur for her special study, and her answer is simple: 'It's cute and it lives a long, long way away from Yale.'

Back in 1966, the lemurs presented something of a puzzle. They lacked monkey-like intelligence – they were as daft as brushes – and yet they lived in big, monkey-sized communities. It had been thought that a rise in intelligence would lead to group living. The example of the lemurs suggested that things were the other way about: that group living could arise even among dullards; only later would it tend to select for intelligence.

   Alison Jolly's insight made perfect sense to Michael Chance. If
social living drove primates to acquire intelligence, then intelligence
was a response to a single, over-riding aspect of social living: the
terror of it.

   In a primate community, everyone has to behave correctly towards
everyone else. Age, sex, temperament and position in the heirarchy
all affect how each individual behaves, and how others must behave
back. Some measure of the true horror of this kind of life – which
seems, after fifty million years of evolution, perfectly pleasant, even
desirable to us – can be glimpsed in the wild. Among apes, baboons,
or chimpanzees, most males are hunched and nervous, radiating
insecurity. When a subordinate male Macaque monkey approaches
a dominant male, it takes along a baby to protect itself from an
unprovoked attack. If that sort of behaviour seems merely 'clever',
imagine your reaction if a human were to do the same thing. The
manipulative behaviours that gave us manners, ethics, co-operation
and the rest are the same behaviours that enable hostage-taking,
kidnap and the use of human shields. The bad and the good are
inseparable: they are all, at bottom, behaviours designed to keep one's
head above water in society. We are, to use an outdated term, 'tool-
maker man' – but the first tools we learned to wield were each other.

   Because they are unable to anticipate, moment to moment, the
behaviours of their fellows, primates watch each other all the time.
A primate who wants to get ahead in the troop doesn't necessarily
have to be the strongest, the biggest, or even the most well liked; it
simply needs to attract attention. The primatologist, Jane Goodall,
Alison Jolly's contemporary, came across one male among her chim-
panzee troop who rose to social prominence by pure self-promotion,
by rolling empty oil drums down a hill while whooping and drum-
ming.[5]

   A primate in authority is, by definition, a primate that commands
visual attention. A young female mountain gorilla studied by another
primatologist, Diane Fossey, stared for hours every day into her troop
leader's face. If he tossed a glance her way, she shuddered with pleasure. Among the higher primates, the way unpredictable behaviour
is deployed acquires a Machievellian, 'treat 'em mean and keep 'em
keen' sophistication. The dominant animal's 'subjects' look to their

leader for cues. Moment to moment, they do not know whether they are to be punished or rewarded. They do not know whether to sit or stand. They do not know what their master feels, so – simulating what they hope is their master's mood – they watch attentively for the smallest change of expression.

We instinctively know that a look can wound, and a glance can save. Now we think we know why: it is a power that has been amassed by the eye over the fifty million years we have spent in Hell; if, like Jean Paul Sartre, you incline to the view that Hell is other people.

In a letter to the magazine *New Scientist* in August, 1991, a reader, Harry Miller, recalled taking a party of blind children to London Zoo:

> I had the idea that I might allow the children to feel and cuddle the baby chimps, learning about their hair, hands, toes and so on, by touch. The experiment, however, proved to be a disaster. As soon as the tiny chimps saw the blind children they stared at their eyes – or where their eyes should have been – and immediately went into typical chimpanzee attack postures, their hair standing upright all over their bodies, their huge mobile lips pouting and grimacing, while they jumped up and down on all fours uttering screams and barks that rose in crescendo . . . I ushered out the children mumbling whatever excuses I could think of.[6]

The very casualness of our language is testament to the social importance of the eye. 'His glance just melted me,' we say. 'They looked daggers at me.' 'She shot me one of *those* looks.' When Miller's blind children failed to play by their visual rules, the chimpanzees of London Zoo expressed their affront in ways both shocking and violent.

Of the thirty-plus regions of the primate brain dedicated to vision, several specialise in spotting social signals. In these regions, there are brain cells primed to spot faces. Some of these cells respond most strongly when the head is pointed in a certain direction; others grow excited when someone looks askance, so that the eyes are pointing in a direction different to the head. This battery of gaze detectors may have evolved in response to the way the faces of great apes have flattened over time. As binocular vision became more important and a good sense of smell became superfluous, primates lost their muzzles. Watching where someone's muzzle is pointing is a simple business: tracking a relatively flat face requires more sophistication. Adopting an upright posture would have made gaze detection even more problematic for humans, further loosening the relationship between gaze direction and posture.

A quick glance back at Alfred Yarbus's record of how the eye studies a portrait – on page 18 – reveals how well our faces are designed to reveal our emotions. The eye in Yarbus's study had difficulty investigating evenly illuminated areas, and could only leap from one area of high contrast to another. This meant that the eye was drawn to those very areas of the face which expressed the widest range of emotions: the lips, the eyes, and the eyebrows.

Armed with gaze detectors, and displaying strong visual markings – eyebrows, high cheekbones and bright white sclera – humans are ideally equipped to develop a further, even more sophisticated survival tool, that takes social thinking to a new and exciting level. This is empathy: the knowledge that others operate in much the same way that you do. If you are looking, they are looking. If you can see the ball, they can see the ball. If you show an interest in the banana, they, too, will be drawn to the banana.

Two-month-old human babies pay more attention to the eyes than to any other part of their mother's face. Six-month-old children spend two to three times as long looking at faces that are looking at them. From about eighteen months of age, human children, engaged in some activity or other, begin checking on the level of attention they are arousing in others. A glance which moves from an object to a parent's face and back to the object asks a question: 'Are you seeing this?' It also prompts a response, as the adult's attention is drawn by the child's glance. Thus reinforced, checking behaviour rapidly develops into pointing behaviour. The question, 'Are you seeing this?' becomes a request, 'Look at this!' The way in which we recruit each other's attention is central to the way we learn language as infants. 'Banana!' I might announce, pointing to a banana, but if you don't know what the pointing is for, you'll never know what a 'banana' is.

Although we are undisputed masters of this behaviour, it is by no means unique to humans. Baboons use quick, furtive glances to recruit helpers when they're being challenged by another baboons, and regularly fool each other by pretending to spot non-existent threats. Whether any primates other than humans really appreciate that their fellows have internal lives is open to question, because no one has quite worked out an experiment that would reveal the answer, one way or another.

It is hard to say when human children acquire this knowledge. Simon Baron-Cohen of Cambridge University believes this understanding is not innate, but develops over time, prompted and reinforced by the pattern of back-and-forth glances that pass between the child and others. 'I think that long before kids are interested in trying to manipulate your attention, or follow your direction of attention, they are fascinated by the eyes,' he says. 'A lot of their behaviour may not be mentalistic.'[7]

The experience of people with autism seems to support this. Autistic people understand 'the facts of vision'. They know how to hide something from another person's view. They understand the rules of geometry; that an object can be visible to one person while being hidden from another. But in spite of this, they have difficulty equating gaze-direction with seeing. They are unable to model another person's internal life; they cannot therefore know that another person actually experiences what he or she sees. A baby of twelve to sixteen months knows to check that its mother is looking at it before it points. Autistic children fail to check, if indeed they point at all, and some, seeing another person point, will miss the meaning of the gesture altogether and focus on the tip of the extended finger.

From the behaviour of autistic people, we learn something important, and heart-rending: being human is a skill we teach each other; and we teach it, first of all, through our eyes.

We do not merely look at faces. We read them. An adult primate's facial expressions will guide the behaviour of its offspring, especially if the offspring is nervous or uncertain. (Next time you see a small child freeze at the top of a playground slide, see if you can spot its anxious-looking parent.)

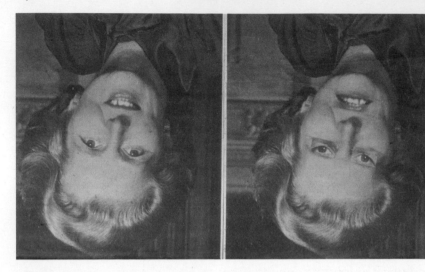

Here is an inverted picture of the former British prime minister Margaret Thatcher. Even though the picture is inverted, it's clear enough that she is smiling. Now turn the book around: ghastly, isn't it? Thatcher's mouth is stretched into a zombie grimace, and her eyes are the eyes of a corpse.

We have Peter Thompson of York University to thank for this chilling illusion. In 1980, Thompson manipulated this photograph by cutting out her mouth and eyes and pasting them back on the picture the wrong way round.[8] You didn't take much notice of that when you looked at the inverted face, because the meaning was clear enough: you saw a smiling mouth, the tell-tale creases at the corners of the eyes, and were content. When you turned the book around, the head was the right way up, but now the former prime minister's *expression* was inverted. With its meaning stripped away, the human face is monstrous.

Paul Ekman, a consultant to both the Dalai Lama and parts of the US Defense Department, is an acknowledged expert in reading faces. He is very careful about who he works with. Gorbachev wanted him to train his secret service; he declined. A US president wanted to learn how better to mislead his adversaries in summit meetings, and recieved the same answer. 'I do not teach people how to be better liars,' Ekman says. The increasing technological anonymity of human society is doing a fine job of that already, without his help.

Ekman was brought up in Newark, New Jersey; at school, he was a year ahead of the novelist Philip Roth. Thrown out of high school for talking back to his teachers, the fifteen-year-old Ekman entered the University of Chicago. There, he discovered Freud. 'I could quote Freud on any topic you might mention,' he recalled in interview.[9] 'It must have been pretty obnoxious.' Convinced that psychotherapy was a cure for the world's ills – a popular view in the fifties – Ekman sat in on psychotherapy sessions. 'I was impressed that there was more than the words going on. So I thought for my research . . . I would look at what people did with their face and body. Little did I realize that that would then occupy my whole life, and I'd never get back to psychotherapy.'

At the beginning, Ekman steered clear of the face. It was too complicated, and no one quite knew how to deal with it. No one, that is, except for Silvan Tomkins, a philosopher more than a psychologist, mistrusted by his colleagues because he was a theorist at a time when, in Ekman's words, 'psychology didn't like theoretical people.' Brought together by the editor of an academic journal, Tomkins and Ekman discussed the central problems of non-verbal communication. Ekman had been given a grant to find out what was universal and what was culture-specific about expression and gesture. Everyone seemed to have an opinion, but nobody had managed to amass any evidence. The celebrated anthropologist Margaret Mead believed that facial expression was culture-specific. Others, including Silvan Tomkins, disagreed; they believed facial expressions were universal and innate, and they fielded an especially heavy hitter: Charles Darwin.

In his speculative book, *The Expressions of Emotions in Men and Animals*, published in 1872, Darwin wrote: 'It has often struck me as a curious fact that so many shades of expression are instantly recognised without any conscious process of analysis on our part. No one, I believe, can clearly describe a sullen or a sly expression; yet many observers are unanimous that these expressions can be recognised in the various races of man.'

Darwin had plenty of circumstantial evidence. He had, for example, noticed how his six-month-old son would respond to his nurse's show of mock sadness by turning down the corners of his mouth.

Far too young to have acquired the concept of sorrow, the baby already knew how to express it. (Subsequent studies have shown that babies of four to seven months old can already tell the difference between 'happy', 'sad', and 'surprised'.) Darwin, then in his early sixties, was plagued with illness, and unable to travel to confirm his intuition. Instead, he relied on far-flung correspondents, including John Scott, curator of the Botanic Gardens in Calcutta, a remarkable observer who came closest to confirming Darwin's conjecture.

Ekman was no anthropologist; to confirm whether or not Tomkins and Darwin were right, Ekman phoned a neurologist, Carleton Gajdusek, who had been in New Guinea studying a disease called kuru, a human form of bovine spongiform encephalopathy (BSE), notorious for being spread by the consumption of human flesh. Because kuru had such a long incubation period – ten years or so – Gajdusek had to invent novel ways of monitoring the disease's progress. The solution he came up with was film – nineteen miles of it, a medical record astounding enough to win him the Nobel prize, and which captured, along the way, the day-to-day lives of two isolated Stone Age peoples; two contrasting ways of life that, within a couple of years, would be gone forever. Ekman is, to this day, the only person ever to watch the lot. (Not even Gajdusek could bring himself to watch it all.) It took him a year: and in all that time, Ekman never saw an expression he hadn't seen before. There was nothing new.

Convinced that Tomkins and Darwin were right – that facial expressions were universal and not merely cultural products – Ekman then had to produce the evidence. The films were not enough. He was going to have to go out and study these people himself. It was not a prospect he altogether savoured. Although one of the tribes was a peace-loving bunch, the other was very violent. When Tomkins came to see one of Ekman's movie shows, he could tell immediately which was the violent tribe, just by the looks on people's faces.

Ekman took along his friend and long-term collaborator Wallace Friesen, and together they set out into the south-eastern Highlands of New Guinea. They met the peace-loving Fore people – and stayed put. 'The violent one was too violent, and I got too scared to go in there,' Ekman admits.

With them, Ekman and Freisen brought stories and photographs. They sat with amenable groups of Fore tribespeople and told them stories. Then they showed them the pictures. The pictures were of people the Fore had never seen, Berkeley students, expressing a range of feelings: happiness, sadness, shock, fear, envy – the full gamut of human emotion (or at least, as much of it as the researchers could bear to carry with them; humans are capable of making around 7,000 emotional expressions).

Ekman and Freisen wanted to see if the Fore would be able to recognise 'foreign' displays of emotion. They could, and did – although cultural differences did make some small differences in how emotions were interpreted. For example, the Fore had a hard time distinguishing surprise from fear – presumably because their culture was not particularly geared to delivering as many pleasant surprises as Ekman's was. None the less, Darwin's conjecture was conclusively proven. Every human is equipped with a 7,000-'word' emotional vocabulary, and a syntax that offers multiple means of expressing the same emotion. Trivial local differences turn up now and again, but the differences are as nothing, when compared to the similarities.

In 1964, the Finnish psychologist Tapio Nummenmaa published *The Language of the Face*, in which he asked people to say, from cut-out pictures of eyes and mouths, what emotions were being expressed. He found that simple emotions, like sadness or happiness, could be read from the mouth alone, but complex emotions (for example, surprise and frustration) required additional cues from the eyes.

The human eye is built to be noticed, and gaze direction can itself be used to convey emotional meaning. The lateral rectus eye muscle is labelled 'amatoris' in early anatomies, because lovers use it to direct their flirtatious glances. A downward gaze can indicate sadness; looking down and away suggests shame and guilt; looking away is a reliable sign of frustration or disgust. Simply widening the eye can, in conjunction with other facial movements, express everything from shock to arousal to doubt.

The expressions we read on each other's faces are of two basic types. First, there are the gestural cues which we use, half-consciously, to clarify our spoken conversations. These movements of the head and brows, jaw and lips, don't necessarily express how we feel; they simply emphasise what we say. They are a form of punctuation. The second type of expression conveys emotion – and how. Throughout our lives, expressions of emotion instantly reflect our ever-changing moods, automatically, and truthfully.

The problem is, it's hard to separate the emphasis system from the emotive system. This happens whenever we watch Woody Allen. Among the director's many gifts is a talent for raising the inner corners of his brows when he talks. He does this to stress his speech, the way you and I would frown or nod. I cannot do what he does, and chances are you can't either. For most of us, the muscle responsible for this inward movement of the brows is not under our conscious control. It occurs only when we are exhibiting genuine sadness.

Using a 'sadness' cue to punctuate his conversation gives Woody Allen either an endearing vulnerability, or a saccharine earnestness, depending on your point of view. Either way, Allen cannot be held responsible: he is simply, like the rest of us, using the tools he has at hand to communicate.

Because the face is capable of expressing two things at once – how we feel, and what we mean – we should perhaps be reconciled to the false smiles, manufactured looks of interest, and all the other deceitful gestures of which our eyes and faces are capable. Deceit is often a necessary part of communication; we conceal what we feel to make room for what we mean. The false smile is particularly handy, since we use it most when we are genuinely happy. From an evolutionary standpoint, signalling enjoyment has never been very important; it takes a relatively large amount of sheer joy to make us smile involuntarily.

Our eyes reveal our inner state, whether we want them to or not. It is impossible to manipulate our rate of blinking for any length of time, and the way our irises dilate is quite outside our control. This is extremely handy when we're looking for arousal in a potential mate: when aroused, the eyes blink more often, and the iris dilates. But though the eye tells the truth, even a human eye, painted like a target, is hard to read at any distance. Lovers may look longingly into each others' eyes, but generally we stand too far away to judge the fine responses of each other's pupils – and who in their right mind would try to watch the rate of someone's blinking? With practice, we might be able to spot these clues, but few of us ever bother because we are far too busy with that other great, and greatly unreliable, form of human communication: language. Encoding words in a coherent fashion and decoding them again is a tough task; it takes most of our attention, and we tend to miss much of the nonverbal information our faces express.

It is perhaps not surprising that we have therefore evolved a sort of truce flag – a unique, highly visible method of telling the absolute God-given truth in a way that gets us believed: we cry. Crying is very difficult to fake. Even actors have to generate some feeling before they cry. Reflecting on this, Marc Hauser, an evolutionary psychologist and a professor at Harvard University, was put in mind of an idea that the Israeli evolutionary biologist Amotz Zahavi proposed many years ago, that you can infer the honesty of a social signal by measuring the cost of the expression. Hauser, applying this principle to the eye, regards tears as the human equivalent of a dog rolling belly-up to show submission. 'Unlike all of the other emotional expressions, tearing as an emotional expression is the only one that leaves a long-term physical trace,' he says. 'It blurs one's vision, therefore it's costly.'[10]

Humans have been pack animals for most of their history. Their survival depended upon co-operation. In such circumstances, there is plenty of opportunity for petty deceit, but very little advantage to blatant lying. In a tight-knit community, a truly pathological liar stands to lose far too much if he or she gets caught out.

Yes, the eyes are the windows on the soul, but we don't count each other's blinks, and we don't press our faces close, to watch each other's pupils wax and wane. We don't have to. Liars get found out quite quickly without all that. 'Trusting others is not only required, but it makes life easier to live,' says Paul Ekman. 'It is only the paranoid who forgoes such peace of mind.'

SIX

Theories of vision

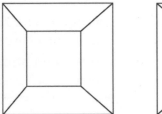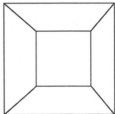

1 – Theon's arrow

At every stage, we will have to reimagine the universe . . .
                                        Arthur Zajonc[1]

Up here in the study, the wind stirs the leaves in the tree outside my window. I am writing this in long-hand and my neglected computer screen dances to the weird, aquatic rhythm of my favourite screen saver. These things do not distract me, nor does the sight of my pen, inching its way across the paper. Nor do the one hundred and one other objects in my field of view. My eyes are much better at filtering out distractions than my ears are. (Try as I might, I cannot listen to music while I work.) In comparison to sight, hearing, touch, taste and smell are relatively passive senses. They are not nearly so capable of focusing on one particular input, filtering out the junk and white noise of daily living.

This intuition seems confirmed and redoubled when you consider that we see far more than we hear, smell, or touch. We are deluged by the visual. The other senses have it easy: we only hear what makes a sound, we only smell what gives off a scent; we only touch what makes physical contact.

Perhaps that is why our vision is so good at concentrating on the thing at hand; at filtering out the noise. The eye cannot afford passively to drink in the scene before it. We would drown in the chaos. It must select. It must hunt.

Theon of Alexandria, considering the shape and movements of the eye, saw clearly the difference there must be between vision and the other senses. The ear, for example, is shaped to receive the air which carries sounds. Clearly, ears are collecting devices. Eyes, on the other hand, are spherical, mobile; they are not passive gatherers, but hunters. Theon is writing in the fourth century AD in the great Library of Alexandria, a not-especially-gifted mathematician preparing simplified lecture notes for his considerably less gifted students. The heyday of the library is long past, and his death, in 405, will save him from the knowledge of the brutal assassination of his daughter, Hypatia, fated to be Alexandria's last head librarian.

Theon's argument is that form follows function. When we look at the eye in action, it is clear that the eye is not merely 'receiving airs', like the ear or the nose. It is hunting. Though we cannot see them, therefore, the eye must see by emitting a 'dart' of some sort or another. Theon is no eccentric. His arguments are bolstering an idea of vision which has pertained for over eight hundred years, and will continue to shape thoughts about vision for at least another millennium.

Why should the eye emit 'darts'? Why would anyone cook up such a bizarre fiction, simply to explain vision? Isn't it obvious that light enters the eye?

It isn't. The idea that the eye sees by capturing light is so counter-intuitive that it was only accepted – and only in the broadest detail – a mere five hundred years ago.

Arthur Zajonc, professor of physics at Amherst College, Massachusetts and the author of a handsome history of light, *Catching the Light*, has invented an apparatus specifically to demonstrate that the subject of his life's work isn't actually there. He calls it, with nice irony, his light box.

It is, as the name implies, a box, filled with light, except that Zajonc has been careful to ensure none of it actually reflects off the sides of the box. Standing beside it, you can see the lamp burning merrily away, and the hole where the light enters the box, and then, peering into the box through a peephole, you see – nothing.

Nothing. An absolute blackness.

You can push a wand through the side of the box, and in doing this you can see – the wand. It is brilliantly lit, but it's only the wand you are seeing. Not the light. The box is full of light, and yet you cannot see it.

We see objects. We see bright objects, and dark objects. The sun is very bright. A piece of coal in the shade is dark. In the sunlight, the coal glitters; it even shines. These observations do not of themselves imply any connection between vision and light. Light is a property of things. We see bright things, not brightness.

For the early Greek philosophers of the atomist school, vision was the ability of the eye to see objects. Visual impressions do not arrive muddled up in a sensory soup, as sounds and smells do. Entering a room, I see a table, a chair, a bed. I do not need to sort these objects out. Even though the chair is behind the table, I perceive, from the little I can see of it, that the chair is a chair. Neither do I struggle to turn my attention from one object to another. Leucippus, and the atomist tradition he founded – a tradition that would last, with many revisions and controversies, for five hundred years – based his theory on what is still the most important phenomenon of vision: its relationship to attention. We see objects, not splodges, or waves, or impressions.

Armed with nothing but an appreciation of matter and a notion of how things are arranged in space, ignorant of optics, ignorant of anatomy, operating mostly without mathematics, the atomists promulgated a daring explanation of vision: that objects release thin films, the way a snake sheds its skin, and these films (or eidola), enter the eye and are apprehended directly by the mind.

Calling upon nothing more than idea of matter in motion, the theory of eidola explains the key phenomenon of vision – our ability to recognise objects even as we see them, whatever the angle, without any effort of interpretation. While we might hear a dog bark, smell a dog's presence, or feel a dog's muzzle against our hand, and by those tokens infer the presence of a dog, only when we see the dog do we apprehend it, directly, without inference. The dog's eidola stream from the dog into our eyes, so that the dog is, quite literally, making its presence felt.

The theory works on several levels. It is, at its simplest, a physical explanation, with implications for the structure and arrangement of matter. For the atomists, all matter was an arrangement of particles. Depending on the density of their packing, these particles are more or less free to move. In the era before optics, mathematics or any theory of energy or radiation, the movement of one kind of material through another – the passage of eidola through the air, or through the jelly of the eye – is a significant idea.

On another level, the eidola theory has considerable psychological appeal, explaining at one fell swoop the whole mysterious business of visual apprehension. That chair in the corner – it has no physical effect that I am aware of. It does not squeak, smell, or brush up against me. It is mute, separate from me. How do I know it exists? How do I know – even more remarkably, how do I know at this distance – that it is a chair? *Because its eidolon is swallowed by my eye.*

The eidola theory hypothesises that vision is, after all, the consequence of an actual physical contact. Everything, in atomist philosophy, is an object. To exist is to exist as an object, and it is in the nature of every object to throw off images of itself. We recognise things, not by any act of memory or comparison or analogy, but because the things themselves – or at least their atom-thin shells – integrate themselves physically into consciousness.

But how could the atom-thin film of a mountain be expected to fit through the tiny aperture of an eye? Come to think of it, how could (say) ten thousand such films, all made of the same stuff, pass without interference, all at one moment, from one mountain into the eyes of ten thousand observers? Even if this were possible – why do objects not run out of mass from which to generate their eidola? Was the universe shrinking? Did objects grow, like snakes, to make up the mass lost in throwing off image films? And so on – a catalogue of awkward questions.

A purely material theory of sight violates what we know about how materials behave; the atomists, in trying to get around this problem, were doomed to invent ever more arcane and unlikely theories of matter. Throwing matter into confusion, simply to explain vision, is just not reasonable. Any competing theory of vision would, as its first requirement, have to treat the rest of the material universe rather more gently than did the atomists.

This theory would, ideally, address the telling handful of cases where vision and apprehension do not arrive hand in hand. Consider, for example, the sorry case of the man who drops a needle on his rug. He knows the needle is lying on his rug. He is looking at his rug. But he cannot see the needle. Why not? According to the eidola theory, image-films of the needle should be streaming into his eye. What prevents them?

While the theory explains wonderfully well why it is that we see objects, it conspicuously fails to explain how we are able to turn our attention from one object to another; how we can focus on one object to the exclusion of another, and how, in the case of the needle lying on the rug, we can miss an object that is unquestionably within our field of view.

Ironically, the body of theory that best addressed these problems was neither innovative, nor handsomely rational; nor did it seem from a known author. As far as we can tell, this competing line of thought is a throwback to philosophy's shamanic past, pre-dating atomist theories by several hundreds of years, marking almost a return to the beautiful, intransigent dogmas of Ra the sun god.

For the Nile-dwelling worshippers of Ra, living in the New Kingdom of Egypt around 1500 BC, light was Ra's act of witness. Light was the god's sight, and things existed because Ra saw them. Ra sees – and all is illumined. What happens if we humanise this dogma – strip the divinity out and place a bare forked human being in its place? This is precisely what some of the poetry and philosophy of fifth century BC Greece achieved; the result is the idea that *every* eye emits some sort of 'visual ray' in order to see.

According to Empedocles, the Sicilian shaman and poet who lived from around 492 to 432 BC, the eye actively observes the world, illuminating it with its own spirit, 'as when a man, thinking to make an excursion through the night, prepares a lantern.'

The idea of visual rays, or 'extramission', became a coherent theory for the first time in the writings of Plato, a disillusioned Athenian politician who sought the consolations of philosophy around 399 BC.

Plato integrated the idea of extramission into a philosophical scheme that is a strange mixture of political and esoteric impulses. Since, under Plato's prescription, objects are not 'making themselves felt' as eidola, Plato had to explain how we are able to recognise things. I have not seen this particular chair before, so how do I know it is a chair? Plato's explanation presupposed the existence of universal Forms – metaphysical archetypes to which all objects in the universe at least roughly correspond. The mind compares the image it has received to the metaphysical library of Forms.

If this explanation seems ugly (and it is) we ought, anyway, to concede that Plato's Forms explain why two differently coloured, differently sized and generally rather dissimilar objects can both be apples; how a cushioned couch and a stone bench can both be seats; how a palace and a hovel can both be houses. We group similar items together, and we recognise familiar forms more quickly than unfamiliar ones. The atomists had no explanation for this; Plato did.

I have to admit that I have a partisan preference for the atomists over Plato. If the atomists stand accused of mangling matter in support of their theories of vision, how much more questionable are the arguments of Plato, who invents and populates an entire realm of ideal being to explain the same thing.

But this is the conventional prejudice of a twenty-first century materialist. Plato's world was very different from ours. Matter has always been around to be tested, manipulated and interrogated. The mysteries of cognition are much less obliging; and if sometimes we have a sense of there being categories 'out there' against which we index the properties of material things – think of terms like 'redness', 'roughness', 'goodness', 'monarchy', 'love' – who are we to deny them a place in theory? We live in a materialist age, so we tend to assume that all theories have to be materialist. They don't, and we certainly shouldn't expect any thinker in 400 BC to have denied or marginalised their spiritual sense, or discounted the possibility of other, as yet unbreached, unexplored realms of being.

With its main claim routed, it is inevitable that the theory of eidola collapsed before the theory of extramission. Extramission explains why our hapless chap can't find his needle: the visual ray emitted by his eye has to hit the needle first. It explains how we can move our attention from one object to another: the eye's ray is narrow, taking in one object at a time. It explains why we clearly see just a tiny part of the visual scene, while the rest is a blur: only that part of the visual ray reflected directly back into the eye is strong enough to be perceived properly. It even explains why we can see better in daylight than at night: by inventing a visual 'ray', the theory can presume to invent, for that ray, any properties it wishes; chiefly, in Plato's theory, that it creates 'a single homogenous body' with daylight that makes material contact with the object seen. A reduction in light means a reduction of the medium with which the visual ray can coalesce, and this, in turn, means a reduction in vision. Nocturnal animals presumably emit more powerful visual rays. Indeed, nocturnal eyes positively shine. Can you not see the visual fire burning in a cat's eyes at night?

So we come to the secret of extramission's robustness; both folk wisdom and common sense support it. First, consider the way the eyes of wolves and owls and other nocturnal creatures shine at night: what can this be other than the visual ray lighting their path, 'as when a man prepares a lantern'? Second, Theon's argument from form: does the eye not look as though it is active, while all the other organs of sense are so obviously passive? Finally – and this is Theon's clincher – we might ask ourselves why, when we try to make something out in the distance, do we screw up our eyes? Why all that muscular effort, if we weren't doing something? If images merely entered, wouldn't it make more sense to widen the eye?

Extramission's chief failing is, of course, that ray: how can sight, that most immediate of senses, rely on such an exhausting-sounding physical effort?

Aristotle, a former pupil at Plato's academy, was not at all convinced 'that the ray of vision reaches as far as the stars, or goes to a certain point and there coalesces with the object, as some think'. This is an argument from personal incredulity: *surely* the eye's rays can't reach the stars, 'coalesce' with them, and return with news of their nature, at a glance? These sorts of arguments deserve to have big health warnings chained to them; the world, after all, is full of unlikely wonders.

Happily, Aristotle knew this, and thought more deeply. He observed that nothing is visible unless it is lit, and that all light comes from fire. (The 'cold fire' contained in luminous materials was a provoking curio for Aristotle and later Aristotelians, well into the Western Renaissance – but it was not one that seriously challenged Aristotle's homely observation.) Fire, therefore, makes things visible.

Not everything is visible at once. Objects get in the way and occlude each other. Aristotle's ruse was to suppose that this is equally true of the air; that it too hides objects from the viewer, *unless it is lit*. For Aristotle, transparent objects – air included – were a special class of object that reacted to light. If this 'state change' – from opaque to transparent – occurs in all parts of an object all at once, we no longer have to worry about how long a 'visual ray' takes to leave the eye, coalesce (somehow) with the object in view, and return (somehow) with (some sort of) visual information about what it struck.

Aristotle's achievement was considerable. He explained how even distant objects are perceived instantly, and why objects are visible by day and not by night, All without recourse to the visual ray.

In dropping the ray, however, Aristotle had to throw away our ability to explain how we focus on objects. If there is no ray, and we simply receive visual impulses, why don't all the objects in our field of view cry out for our attention all at once? We are back where we started.

Around this time – 300 BC – the study of vision began to fracture. It had proved intractable. No one, however gifted or imaginative, was able to come up with a single explanation for all the different phenomena of vision.

2 – Straight lines

Science is facts; just as houses are made of stones, so is
science made of facts; but a pile of stones is not a house and
a collection of facts is not necessarily science.

Henri Poincaré

Twenty-odd years after Aristotle's death, Euclid, a committed Platonist and a prominent Alexandrian mathematician, wrote his own theory of vision. Euclid's *Optica* contains not a single line about the mechanisms of the eye, nor about the experience – the psychology – of seeing. Euclid's interest in vision was purely mathematical.

He was one of the founding fathers of geometry; in the visual ray he spotted an elegant, real-world demonstration of the behaviour of rays and angles. In *Optica*, Euclid developed a simple optics of the sort that today any schoolchild would recognise. So it is very hard for us to appreciate quite how revolutionary this was. For the first time, a visual theory was being expressed in purely mathematical terms. For the first time, vision was being treated as a purely geometrical problem.

Ironically, for a tradition that would prosper well into the last century, we know frustratingly little about the early days of optics and virtually nothing about the motivations and assumptions of its first thinkers. Was mathematics to provide a full explanation of vision? Was it in competition with other traditions? Or did optics start out as a sideline – a neat real-world example of how objects relate geometrically in space? We do not know. Nothing about Euclid's early life is certain. We do not even know the year of his death.

A hundred and fifty-odd years after Euclid's *Optica*, Claudius Ptolemy, another Alexandrian, developed and expanded the mathematics of his day in a wealth of writings, most of which have been preserved. Sadly, among the few books that are missing are the very ones in which Ptolemy discusses the physics, psychology and philosophy of vision. Even his optics fares poorly, surviving only in an incoherent translation by the Byzantine scholar Admiral Eugene of Sicily. (Eugene is not entirely at fault; the original was lost, and he was having to work from a rather poor Arabic translation.) It is tempting to speculate that these losses contributed to the split between the mathematical tradition that gave birth to optics, and the more psychological, philosophical approach to vision, typified by Plato.

Tempting; but in the end, it was inevitable that optics would dominate vision theory for the next thousand years. Unpicking the role of light in sight was, quite simply, the single most important breakthrough in vision research. While psychology played its part in the breakthrough, the greatest gains in understanding were achieved through the clear, cold proofs of Euclidean geometry.

*   *   *

I am writing this book in wartime. Every other night, another casualty is seen being rushed through the doors of the Al-Kindi General Hospital in Baghdad. One of the more depressing aspects of this project has been the way the news of Iraq's seemingly bottomless slide into dereliction has thrown up the places and names of the Arab Renaissance – a period quite as significant as the Renaissance in Christian Europe – which transmitted, codified, expanded, and improved the intellectual inheritance of over a millennium of Mediterranean culture.

Baghdad's beleaguered general hospital is named after Abu Yusuf Ya'qub ibn Ishaq al-Kindi, scion of the royal house of Kindah, and father of modern medicine and pharmacology. Al-Kindi was born late in the eighth century AD in al-Kufa, where his father was governor. His adult life was spent first in Basra, then Baghdad – the twin poles of the Islamic world.

The intellectual revolution that gripped Europe six hundred years later was driven by artisans, supported and patronised by a leisured class of middle-ranking nobility. The Islamic renaissance was of a different order, a cultural project consciously incited and bankrolled by the more-or-less benevolent dictatorship of the Mohammedan Caliphate. Cosmopolitan and aspirational, the empire had to reach beyond its own traditions and borders to achieve its cultural ambitions, and had relied for the most part on a succession of Christian Syrian doctors to fill its libraries and write up its intellectual achievements. Al-Kindi's soubriquet 'the Philosopher of the Arabs' not only saluted his exceptional grip on logic, geometry, mathematics, music and astrology; it also acknowledged that he was the first native Arab thinker to make much impression on the world.

Though Al-Kindi was the Caliphate's first native great thinker, he seems to have been immune to the temptations of chauvinism. 'We ought not to be embarrassed of appreciating the truth and of obtaining it wherever it comes from,' he wrote, 'even if it comes from races distant and nations different from us. Nothing should be dearer to the seeker of truth than the truth itself, and there is no deterioration of the truth, nor belittling either of one who speaks it or conveys it.'[2]

Famously avaricious, Al-Kindi spent his most productive years in anonymous seclusion in an affluent business suburb of Baghdad, caring more for his garden and private zoo than for his neighbours. Inevitably, he became a target for envy.

The Banu Musa brothers – Jafar, Ahmad, and Al-Hasan – were recruited to Baghdad's House of Wisdom at the same time as Al Kindi by the Caliph Al-Ma'mun. Al-Ma'mun was building up the greatest library since the Alexandrian; translators were in high demand, particularly those with the expertise to unpick Greek scientific texts.

We don't know precisely what sparked the enmity between Al-Kindi and the brothers, though there are some intriguing hints: the degree to which Al-Kindi could not be bothered to translate his own Greek texts, but preferred to adapt the translations of others: the scale and significance of the private collection of manuscripts he amassed while working for the library. Neither do we know how long the brothers' dislike festered before the death of Al-Ma'mun, followed by that of his brother and successor Al-Mu'tasim, left the House of Wisdom in political turmoil. When the despot Al'Mutawakkil succeeded to the Caliphate in 847, the brothers showed their hand. Ingratiating themselves with the new Caliph, they saw to it that Al-Kindi was beaten and his extensive library confiscated.

The old cosmopolitanism was over, and it is possible, given Al-Kindi's lack of interest in religious argument, that Al-Mutawakkil's maltreatment was of a piece with his persecution of his non-orthodox and non-Muslim subjects. Al-Kindi lived the rest of his life in a sadly reduced city, its vistas marred here and there by the ruins of churches and devastated synagogues.

Al-Kindi's optical work is an impressive, if confused, attempt to re-bind the physics, geometry and psychology of vision, with geometry for the first time carrying by far the greater part of the argument. Al-Kindi, the good geometer, was an extramissionist. His arguments for the existence of a visual ray were predicated on the fact that we see most clearly in the central part of our field of view, and rather fuzzily elsewhere. His discussion of this point is interesting, in that it is the first clear instance we have of someone using reading to elucidate the behaviour of the eye. If the eye simply 'received impressions' passively, Al-Kindi argued, there would be no need for the eye to jump from word to word on a page, for the whole page is in view all the time one is reading. The eye's movement along a line of text suggested, to Al-Kindi, that the eye was bringing something to bear on the page – a narrow cone of what we would call attention, but which for Al-Kindi was a real, if invisible, ray.

Al-Kindi's greatest single contribution to vision research was, ironically enough, a mistake, a happy inconsistency between his views on the behaviour of the visual ray and an important insight he had into the behaviour of light. Al-Kindi remarked that light pays no heed to the objects off which it reflects. The *form* of the object does not influence light's behaviour. Light simply rebounds off every point of every surface, at every angle.

This is, essentially, a philosophy of radiation, in which everything affects everything else, all the time, through the transmission of power.

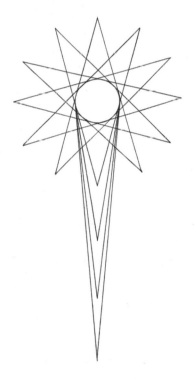

Al-Kindi's theory of light as radiation was neatly captured, six hundred years later, in this figure by Leonardo da Vinci. If light is scattered off an object at every point, in every direction, then an infinite number of focused images of the object will be available to observers regardless of their location or distance.

Gone, in Al-Kindi's philosophy, is the transactional view of the world, promulgated by the atomists, where objects send exquisitely scaled-down material messengers, like picture postcards, into the individual observer's eye.

When Al-Kindi applied this observation to vision, all went well at first. He assumed that all radiations behave the same. If the visual ray behaves like light, hitting every point of every available surface in a straight line, this serves to explain why we see objects in perspective. Imagine, for example, a wheel. Seen face on, it makes a circle in the eye; lying on the ground, however, it makes an ellipse.

Al-Kindi used the same simple analogy between visual rays and light rays to explain why the central part of our field of view is so much clearer than the periphery. Assuming that the whole surface of the cornea is responsible for vision, he argued that any object directly in front of the eye would be exposed to visual rays emitting from every part of the cornea. An object to the side, however, would receive only a proportion of the rays, and thus it would make a rather poorer mental impression.

This is the problem Al-Kindi failed to spot: if visual rays behave like light rays, they presumably behave like light rays during their return trip to the eye, as well as on their outward trip. In which case, why should they return to exactly their point of origin on the surface of the cornea? What is there to prevent rays from each point of the object from striking every point on the cornea, blurring the view hopelessly?

By introducing the idea that every point of a surface reflects light in every direction, and by making a metaphorical link between light and the visual ray, Al-Kindi opened up a world of trouble for himself, and a world of possibility for subsequent thinkers. Al-Kindi had effectively, if unintentionally, proved the superfluity of the visual ray. Leonardo da Vinci's figure makes this point with particular eloquence: it is light, and light alone, that makes vision possible.

By his mid-forties, Abu 'Ali al-Hasan ibn al-Hasan ibn al-Haythem had reached the end of his tether with his desk job. As minister for Basra, it was his job to mediate between and contain all the religious factions in the city – work for which he felt little enthusiasm and

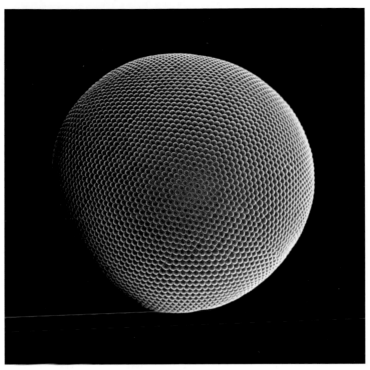

Wraparound vision: the compound eye of the Antarctic krill
Euphausia superba.

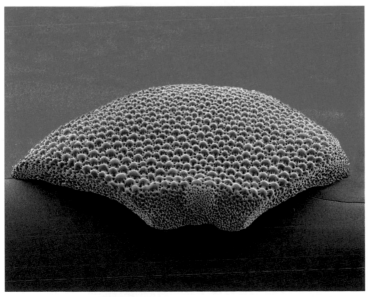

Smothered in tiny crystal eyelets, a brittlestar's carapace doubles
as a primitive compound eye.

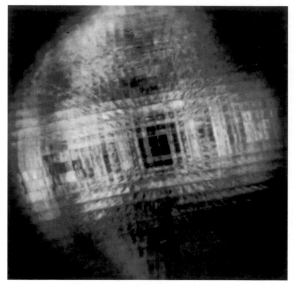

The compound eyes of shrimps, prawns, lobsters and crayfish have mirrored boxes in place of lenses. *Left*: the eye of a decapod shrimp. *Right*: close-up of the surface of a crayfish eye.

Sigmund Exner's photograph – the first experiment of its kind – is focused, not through a lens, but through the optics of the superposition compound eye of a firefly.

The mantis shrimp crams up to sixteen different colour receptors into the horizontal bands running across its eyes.

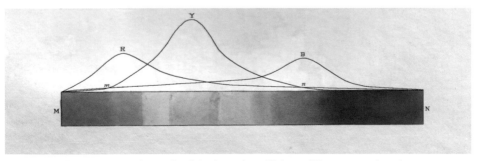

David Brewster's graph of the intensity of light at different wavelengths. Unwittingly, he has revealed the relative sensitivities of human cones.

Black-eyed susans (*Rudbeckia hirta*) photographed in visible light (*left*) and in ultraviolet (*right*). Pollinating insects are often drawn to patterns our eyes are not equipped to see.

The camouflaged eye: the banded butterflyfish *Chaetodon striatus*.

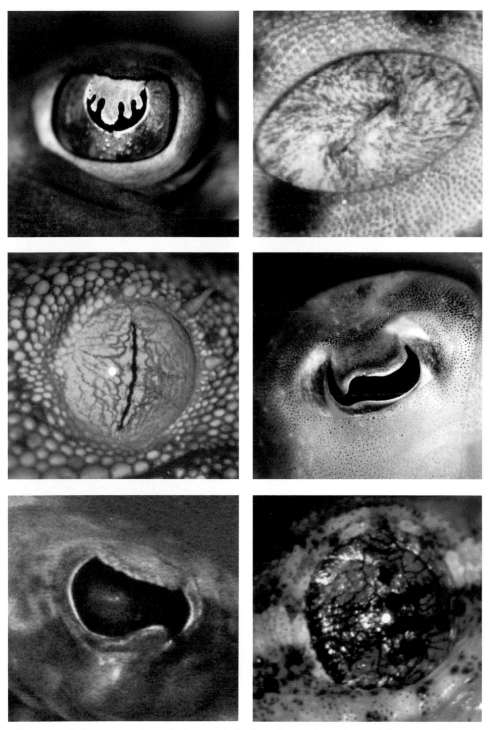

Round pupils like ours make optical sense, but other shapes abound, especially among fish and squid. Camouflage is part of the reason, but there may be undiscovered visual advantages to each shape. *Clockwise from top left*: blue-spotted ray; epaulet shark; harbour cuttlefish; pale gecko; yellow flounder; spiny-tailed gecko.

Novel optics: the sparsely faceted compound eye of a Phacopid trilobite.

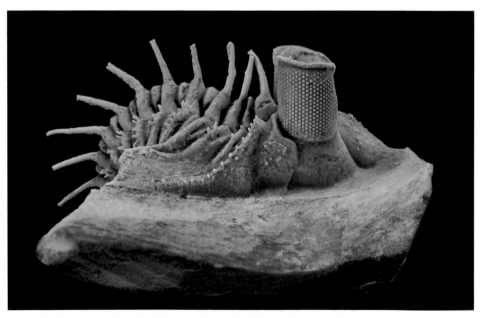

A trilobite with a sunshade: *Erbenochile*, shielded from the glare of its sunlit habitat, enjoyed wraparound vision but, like other trilobites, it could not look up. It did not need to; there were no predators around to swoop down on it.

Vision through touch. *Above*: a star-nosed mole's nose boasts a central 'acute' region, like a fovea. *Right*: a naked molerat acquires spatial information from its sensitive teeth.

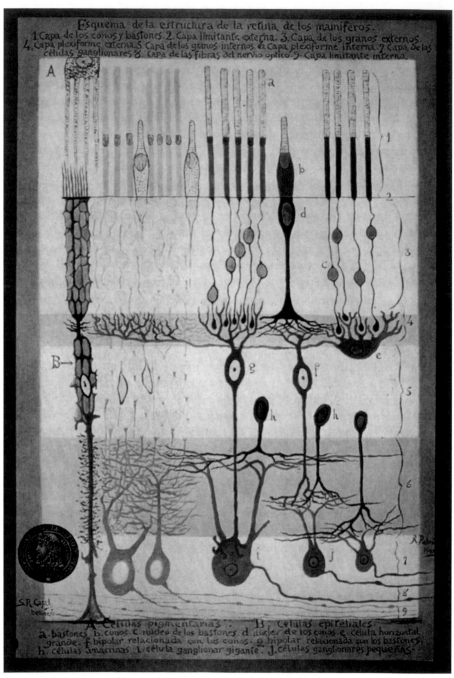

The nerve cells of the retina, drawn by Santiago Ramón y Cajal. Cajal's description of a nervous system made up of distinct, individual cells of different types won him a Nobel Prize in 1906.

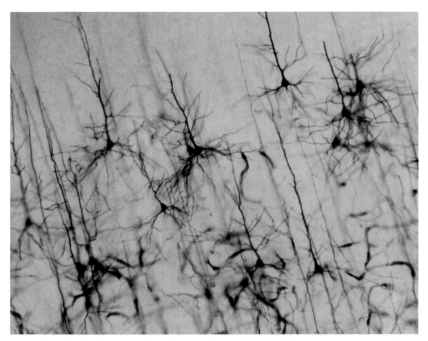

A section through the visual cortex stained by the Golgi method fixes entire nerve cells, but only a tiny fraction of the cells so treated respond to the process. The result is a beautiful, radically simplified image of how individual cells connect to form a single nervous system.

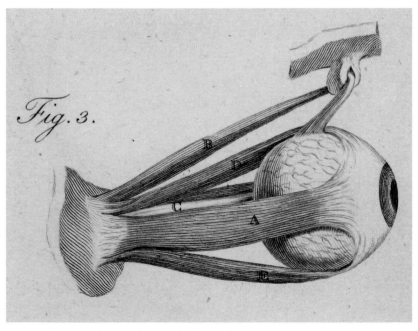

The musculature of the eyeball according to the nineteenth-century American physician David Hosack.

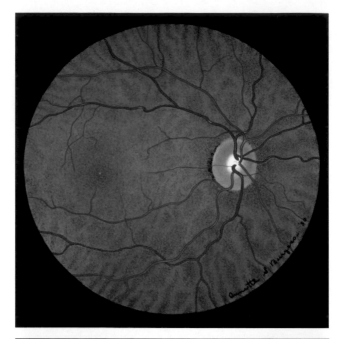

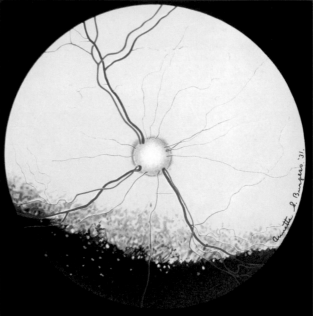

Views through an ophthalmoscope.
Top: the right eye of a brunette Italian woman, 36 years of age.
The fovea is a barely visible yellowish disc, free of blood vessels,
to the left of the nerve head. *Bottom*: the left eye of a grey cat.
These exquisite paintings by Annette Burgess, published in 1934,
capture anatomical detail better than any photograph, and even
record the play of light as it glances off the cat's tapetum.

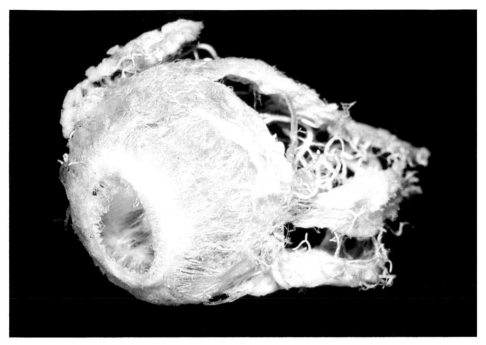

The eyeball is largely made up of tiny blood vessels, as this corrosion casting reveals.

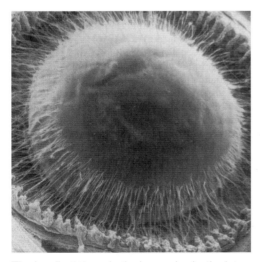

The loneliest tissue in the human body: the lens of an eye, suspended and connected to its ciliary muscle by thousands of tiny elastic 'guy wires'.

A biological sundial: the pineal organ of the desert spiny lizard.

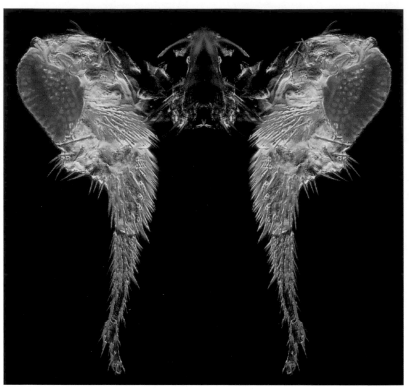

The ectopic eyes on the legs of this fruit fly were formed by artificially activating the fly's *eyeless* gene.

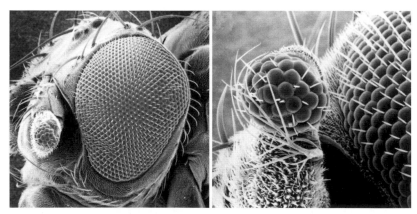

The compound eyes on the antennae of this fruit fly were triggered by the gene *smalleye*, from a mouse.

John Dalton (1766–1844)
portrait by Benjamin R Faulkner.

Thomas Young (1773–1829)
portrait by Henry P Briggs.

Sir Charles Wheatstone (1802–1875)
portrait by Charles Martin.

Santiago Ramón y Cajal (*left*) playing chess in
1898; a photograph taken by one of his children.

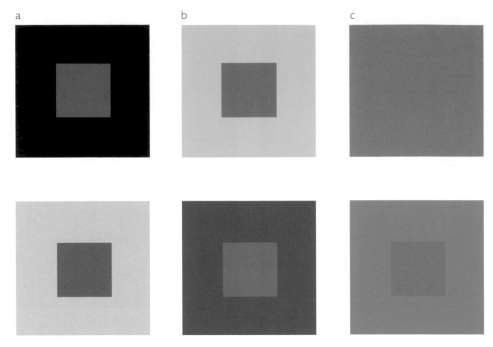

Three contrast effects (the squares in the centre of each pair are identical). Whether detecting brightness (a), hue (b) or intensity (c), the eye responds to contrasts, not absolute values.

Edwin Land prepares an experimental 'mondrian' to test the robustness of colour vision.

a

b

c

Because it exaggerates the boundaries between Ernst Mach's even grey bands, the eye generates an illusion of fluted columns (a). The illusion is almost as strong using bars of differing saturation (b) and hue (c).

Look at it long enough, and this complementary-coloured Union Jack produces a true-colour afterimage. See page 226 for an explanation.

Adelbert Ames's 'full-size monocular distorted room'. An exterior shot reveals the architectural trickery behind the illusion.

less aptitude, coming eventually to the conclusion that truth resided solely in the sciences. Science was Al-Haythem's private passion, and his considerable knowledge was the product of years of private study. Encouraged by his growing scientific reputation and hungry for adventure, Al-Haythem (spelled 'Alhazen' in most European sources) made his bid for freedom and went off, with a party of engineers, to build the Aswan dam.

Today's Aswan dam is three and a half kilometres long, and holds back about 165 square kilometres of water. The reservoir is about 480 kilometres long; approximately the distance from London to Newcastle. Filling it took twelve years and displaced 90,000 people. Al-Haythem and his party of engineers, arriving in the year 1009, had bitten off rather more than they could chew.

The only reasonable thing to do was to go home and admit defeat. There was, however, one obstacle – the Caliph Al-Hakim, Al-Haythem's cruel and neurotic patron, a man who, in true *Mikado* style, had honed his homicidal cruelty on dogs (he disliked their barking, and had them all killed) before working his way up to people. No one remained Al-Hakim's enemy for long.

Al-Haythem contrived to elude Al-Hakim's displeasure by feigning madness. Given the Caliph would occasionally take it upon himself to ban certain vegetables, an alibi of insanity must have seemed quite reasonable. Al-Hakim sighed, wrote off his losses, and consigned Al-Haythem to house arrest.

Al-Haythem did not have to try too hard to appear crazy. After all, he blacked out his windows and spent twelve years conducting experiments in the dark. In 1021, Al-Hakim died. Al-Haythem opened his windows and announced to his bemused guards that he wasn't mad, after all; he'd only been pretending. Incredibly, they believed him.

Al-Haythem was interested, first and foremost, in the geometrical behaviour of light. He developed several beautiful and influential demonstrations of the way light moves in straight lines. One experiment, involving five candles and two rooms connected by a small hole in the common wall, would revolutionise ideas about vision and the eye.

Al-Haythem arranged five candles in an otherwise unlit chamber. Light from the candles entered the second chamber through a small

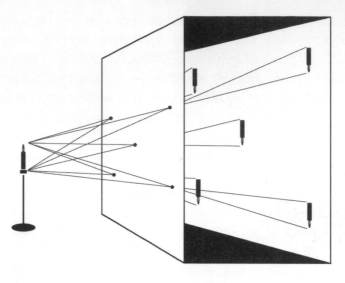

Al-Haythem's optical experiment – the first we know of ever to
use a camera obscura – shows that light moves in straight lines.

aperture in the wall. The result: five clear patches of light formed on
the wall of the second chamber, opposite the aperture. This effect
can only be explained if light moves in straight lines.

Al-Haythem wrote the first really good studies of refraction; he wrote
about sunsets, rainbows and eclipses and was the only writer of the
period to rival the Roman anatomist Galen in his description of
the parts of the eye. He was also the first writer we know of who understood the role of light in vision. Al-Haythem knew that light, when it
travels from a lighter to a denser medium, changes its direction. He
understood the implication of this for curved surfaces: that two parallel
lines, hitting a curved object of greater density, will converge. In fact,
all rays hitting a curved surface will refract, except that ray which hits
the surface at a perfect right angle. That ray alone will not be refracted.

Al-Haythem understood from his anatomical knowledge (as Al-
Kindi did not) that vision takes place inside the eye. Since rays must
pass through the pupil of the eye to be seen, this effectively restricts
the number of angles at which unrefracted rays can penetrate the
eye. Light entering the eye from objects straight ahead will not be
refracted, but light entering the eye from objects to one side will be.
Since the view straight ahead is clearer than the view in the periphery,

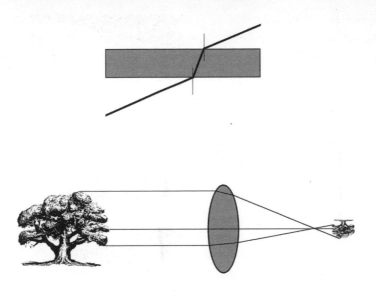

Refraction and focus. *Top*: light entering a denser medium veers
towards the perpendicular.

Bottom: if the denser medium is lens-shaped, the combined
course-changes upon entry and exit draw parallel rays of light to a
point. Beyond this point, images captured by the lens are inverted.

Al-Haythem concluded that unrefracted rays are stronger than
refracted rays.

This was very clever indeed. Al-Haythem had explained the character of the eye's field of view purely in terms of light, without any
recourse to visual rays. This was a tremendous advance, combining
anatomy, mathematics and psychology to explain how we see.

But there was trouble ahead. Al-Haythem, following Galen, understood that the eye is connected to the brain by the optic nerve, and
that the brain is the seat of consciousness. This left him with a serious
problem; if light enters the eye and travels up the optic nerve, what
was the lens, the 'glacial humour', for?

It couldn't be a focusing device, because the eyeball itself is a lens.
Al-Haythem conjectured that light rays, passing through the lens, were
made parallel, so that they might travel up the optic nerve to the brain.

In the absence of good anatomical understanding of the retina, this
is a clever idea, but it is seriously flawed, for it imputes unexplained

properties to the lens. How can a lens take a cone of converging light and make it *diverge*, so that it runs in a tight, parallel beam? There are two pressing reasons why Al-Haythem, at this crucial point in his scheme, abandoned optics in favour of what is, at best, speculation.

First, unless the light passing from the glacial humour runs in a parallel beam, he could not explain how light reaches the brain. He had no theory of electricity or nerves, so direct communication of light to the brain was his only means of explaining how the brain sees.

The second problem is more subtle. Even if Al-Haythem supposed some other, non-optical, means by which the optic nerve could pass visual information to the brain, the image cast on the optic nerve by a regular lens would make no sense. The image would be inverted, just as the candle images are inverted in the drawing on page 168. And this is an obvious nonsense. Inverted images cannot possibly play a role in vision. How could they?

If they did, would we not see the world upside down?

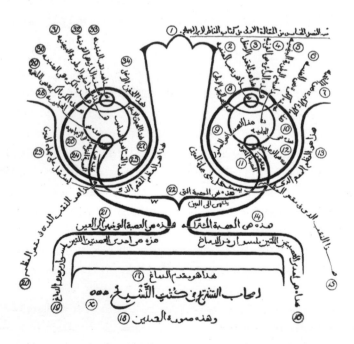

The oldest surviving pictorial representation of the visual system, from Al-Haythem's *Book of Optics*, written around 1038.

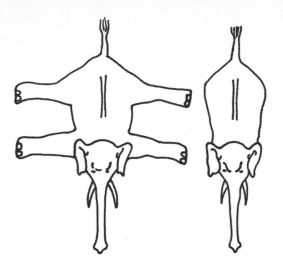

3 – The world turned upside-down

We see nothing, save through reason.
            Schopenhauer, *On Seeing and Colours* (1815)

In 1962, the anthropologist William Hudson showed a group of adults
and children living in the Zambian bush two drawings of an elephant.[3]
In one, the elephant was seen from above; in the other, the same
elephant was squashed, as if by a steam-roller, so that its legs and
trunk were splayed out to the sides of the body. The children preferred
the 'squashed' drawing, because it contained more of the elephant.

When Westerners are shown the same drawings, they prefer the
unsquashed one. Although there is less elephant in it, they consider
the picture more realistic, since it captures what it would be like to
see an elephant from that particular angle.

Both choices are sophisticated, aesthetic decisions. The pictorial
art of the veldt typically conveys ideas of value and meaning; the
art of the post-Renaissance West typically simulates the rules of
optics. People prefer some representations over others. Representations are plastic, modified over time. They change.

Anyone with the ability, opportunity and inclination to read a book like this is going to be living in a world of representations: a world of signs, moving images, and visual puns. Our world is so rich in visual representations that some philosophers ask whether we are losing touch with the real world entirely.[4]

Images and signs were certainly part of the world in Al-Haythem's time, but they were far from being dominant. The relationship between what was seen and what was really out there seemed, for that reason, perfectly straightforward. The eye apprehended the truth which lay before it. Why make life more complicated?

The trouble with an eye that sees upside down is that it requires the brain to turn things the right way up. The brain cannot simply apprehend what is out there – it has to manipulate the information it receives and then re-present it. But to whom, exactly, is this right-way-round image being represented? Is there a screen in the brain, and a watcher, watching it?

Al-Haythem and his successors winced away from the idea that eyes received inverted images, for the very good reason that for this simple optical quirk to work, a new philosophy of mind was required – a view of perception that would drive a wedge between the world itself and the apprehending mind.

If everything we see is manipulated before we see it, how do we know that anything is true? Suddenly, the world is split in two: the world we sense, and the world represented by the senses. If everything we sense is a representation of what's really out there – well – what's really out there?

We don't know. We can't know, any more than I can know, from looking at the icons on my computer screen, exactly what is going on in the silver box under my desk. I can control the box, as much as it will let me. I can drag files into a 'wastebasket'. I can shuffle files from one 'folder' to another. I can draw pictures. I can type words, delete them, and get them back again at the touch of a button. But I can't understand what's really going on inside this box of tricks, because the icons I click and drag need not in any way resemble what's really going on inside the box.

This vision of the world reduces philosophy and science to a user's manual, a description of how we interact with the world, which says nothing whatsoever about the way the world really works. It's a predicament which Donald Hoffman, an expert in computer vision, recognises and explores with unusual candour in his book *Visual Intelligence*:

> Neither biology nor quantum theory dictates the nature of the relational realm. Nor does any other science. Each studies certain phenomena, and describes these by precise theories. In no case do the phenomena or the theories dictate the nature of the relational realm. We might hope that the theories of science will converge to a true theory of the relational realm. This is the hope of scientific realism. But it is a hope as yet unrealised, and a hope that cannot be proved true.[5]

For six hundred years, the philosophers of Europe's Dark Age fought shy of this demoralising, ultimately unavoidable, conclusion. The best of them was William of Ockham. Born in Surrey just before the turn of the fourteenth century, Ockham bequeathed the European philosophical tradition some powerful weapons. 'Ockham's razor' (often spelled Occam's razor) is the informal-sounding but powerful observation that the simpler an explanation is, the more likely it is to be right. Applying his own stricture to the theories of vision that were slowly penetrating European thought, Ockham argued that the mind apprehends things, at a distance, through the application of the intellect. By imprisoning Plato's realm of perfect Forms within the bounds of the individual skull, Ockham made our ability to recognise, categorise and discriminate a matter of mind, rather than a quasi-mystical condition of nature. This simplifies vision theory immensely; the eye sees patterns which the mind can apprehend.

Generally speaking however, the monasteries and universities, the great seats of learning of medieval Europe, more or less neglected optics, regarding it as a discipline whose philosophical riches were exhausted. The thirteenth-century polymath, Roger Bacon, attempted to reconcile the theories of vision, but the result was indigestible. (His list of sources was useful though, as it included Al-Haythem.)

Outside the academies, the practical applications of optics were growing ever more numerous. The arrival, in the fifteenth century of flat glass mirrors was probably the trigger for the series of artistic experiments which led to the discovery of linear perspective.

The mirror image – a three-dimensional scene, captured on a flat rectangular surface – laid before the artist a challenge hard to resist: the possibility of painting pictures that could, like mirror images, be mistaken for reality.

Since a painting is a still image and cannot shift, as a mirror image can, to accommodate the changing position of the spectator, perspective art requires that we look at it from a particular angle. This was not the case with the Gothic art which preceded it, nor with the school of 'naïve' art in our own day; both, though ostensibly 'representational', arrange their objects higgledy-piggledy on the picture plane.

Much has been written on the pervasive influence of perspective art. For our purposes its main significance is the intellectual climate it encouraged. What more can optics say about vision? Nothing – and this is the point. Leon Battista Alberti (1404–72), writing up the perspective techniques of the artist and optical illusion-maker Filippo Brunelleschi, had this – and this alone – to say about the how the mind apprehends what it sees: 'This dispute is very difficult, and is quite useless for us.'

'We', in this context, were the artisans supplying curios and ornaments to an unprecedentedly wealthy merchant class. Were they aware that their optical games were revealing the secrets of vision more quickly than half a millennium of philosophical bean-counting? No. The writings of the great artists reveal a remarkably narrow field of interest. Albrecht Dürer's *Painter's Manual* of 1525 says nothing about colour. Leonardo da Vinci, who studied linear perspective and drew over one thousand images through the camera obscura, was also one of the finer anatomists of his day – yet he thought that the optic nerve is connected directly to the lens and wrote of, not one, but six methods by which upright images reach the brain – each method more wrong-headed than the last. Even Johannes Kepler – the mathematician and astronomer who provided us, in 1611, with our current model of how the eye uses light – was only incidentally interested in vision. Understanding the eye was, for him, only a stage on the path to understanding the movements of the planets.

*   *   *

Tycho Brahe was the greatest astronomer of Kepler's day. He was also the best-funded. Brahe's uncle had died saving King Frederick II of Denmark from drowning. (Returning from a naval battle with the Swedes, the king contrived to topple off his own drawbridge.) In a gesture of gratitude, in 1576 King Frederick gave his nephew, Tycho, an island and money for an observatory. Lots of money: a maintenance grant of 500 daler per year, revenues and firewood from the manor of Kullagaard, the use of eleven farms near Helsingborg, a canonry of Roskilde cathedral, and the fiefdom of Nordfjord in Norway – altogether about one per cent of the crown revenue.

'My opinion of Tycho is this,' Kepler wrote to Michael Maestlin, his old mathematics teacher, 'he is superlatively rich, but he knows not how to make proper use of it, as is the case with most rich people. Therefore, one must try to wrest his riches from him.'[6] Kepler's ability to turn a phrase is a godsend to biographers. Kepler's mother was, according to his own account, 'small, swarthy, gossiping and quarrelsome, of a bad disposition', while his father was a man 'vicious, inflexible, quarrelsome and doomed to a bad end . . . Saturn in VII made him study gunnery.'

He isn't making any of this up. The Keplers were as mad as a bag of cats. Johannes's aunt was burned as a witch, and we still have the lawyers' bills from the time he defended his mother from the same charge.

While he was teaching mathematics in Graz, Kepler was forced to choose between embracing Catholicism (the Keplers were Lutheran) or being expelled from Austria. Tycho Brahe's timely offer of an assistant's job in Denmark was brokered by their mutual friend, the anatomist Johannes Jessen. For the next year, Kepler and Brahe worked together on the problem of the orbits of the planets. Kepler had to wait until Brahe's death, the following year, before he was able to obtain the great man's jealously guarded astronomical data, stealing it from under the noses of Brahe's acquisitive relatives. Only then was he able to make the astronomical breakthrough for which he is best known, ascribing elliptical paths to the planets as they orbit the sun. While Brahe was alive, Kepler's studies were of a humbler cast. For instance, he was given the task of explaining why the moon shrank suddenly during a solar eclipse.

This nonsensical observation was not a little embarrassing for Tycho

Brahe, whose reputation stood – and stands – on the unerring and unprecedented accuracy of his observations, the very best achieved without telescopes. How could it be that an astronomer capable of measuring one minute of arc – one sixtieth of a degree of the sky – could not get a sensible figure for the size of an object as big as the moon? Assuming that the moon wasn't playing tricks, Brahe's observing equipment had to be introducing the discrepancy. But how?

Kepler's explanation, in a letter he wrote to Maestlin, is typically facetious, more concerned with the fact that he got his pocket picked while he was working ('By Hercules!' he wrote, 'what an expensive eclipse!') than with the details of his observation. Kepler studied the eclipse of 9 September, 1600 in Graz, using a camera obscura mounted on axles – the first occasion we know of where tracking was used in astronomy. The camera obscura allowed direct measurement of the diameter of the moon, projected on the wall opposite the aperture. Sure enough, the diameter of the moon, silhouetted against the sun's corona, was significantly smaller than the diameter of the positive image the moon made when observed through the same equipment at night.

What was going wrong? Kepler's first recourse was to the library, to the works of the thirteenth-century writer, John Pecham. When he could make neither head nor tail of Pecham's 'arcane' theorising, Kepler turned to experiment. Using taut threads to represent rays of light, a book suspended from the ceiling to represent a viewed object, and a hole cut in a table to represent the aperture of the camera obscura (the reactions of his domestic staff are not recorded), Kepler constructed an apparatus to demonstrate, to his own satis-faction, why the camera obscura, though useful, was unreliable.

The camera obscura is unable to bring objects into perfect focus. Images cast on to the back wall of a camera obscura might look crisp, but perfect focus cannot be achieved. This is because light passes into the chamber, not through one ideal focus point, but through an aperture. Light rays from every point of the object pass into the box at many angles, which taken all together make the shape of the aperture. The aperture itself introduces blur, so that light from a full moon, blurring ever so slightly, will make a large image, while light bleeding round the edges of the moon during an eclipse, blurring ever so slightly, will make a smaller one. The aperture can never be small enough to eliminate all blur; and the smaller the aperture, the dimmer the image.

This was, all in all, a rather disturbing conclusion for Kepler to reach. The eye was a kind of camera obscura, at least in the sense that light entered it through a narrow aperture, the pupil. If the behaviour of light through apertures led to inconsistent results, how trustworthy, from an astronomical point of view, was human vision? Tycho Brahe, Kepler's mentor, claimed to have achieved an observational accuracy of one minute of arc. What if he were wrong? Kepler had to establish whether the human eye was reliable, and if it was, how that reliability was achieved. Without that knowledge, the best astronomical data of the day could not be relied on.

Kepler did not have to work in a vacuum. He could turn to any one of the thirty classical works on optics republished in the sixteenth century, including books by Euclid and Al-Haythem. But what finally drove Kepler to his astonishing result was the work of Felix Platter, a man entirely unconcerned with the nature of light.

Felix was the son of Thomas Platter, a highland shepherd and vagabond turned boarding school teacher. He made the most of his father's success. A graduate of the University of Montpelier, by 1560 he was one of Basel's more fashionable doctors, attracting aristocratic and bourgeois clientele from as far afield as southern Alsace, Württemberg, and the Black Forest. Fame came at a cost. On a family trip to Grächen, his father's rural birthplace, Felix woke up one morning to find the girls of the region lined up outside his house with gifts of their own urine, hoping for (and to Felix's great credit getting) a free check-up from the famous doctor from Basel.

Felix Platter's medical interests were shaped by the anatomical advances of the Renaissance. These advances were not so much to do with the increasing acceptability of human dissection (though this was certainly important) as with improvements in printing, which provided anatomists with decent, accurate, contemporary illustrations of previous findings. In 1588, Felix Platter published *De Corporis Humani Structura*, a slim volume containing fifty annotated anatomical plates. Felix's anatomy is conscientious, careful, and for the most part unexceptional – an excellent representative example of the anatomy of his day. In one respect, however, Platter's work was certain to startle the informed reader.

With no knowledge of optics, Platter did not know how important it was that the eye should not see inverted images. Indeed, it is extremely doubtful whether he knew that images *could* be inverted. He was free to observe, as neutrally and as accurately as he could, how the eye was made. With a trained gaze and no received opinions, Platter saw what no one had seen before. He saw the crystalline humour of the eye, just behind the iris, towards the front of the eyeball. He saw ciliary fibres holding the crystalline humour before the pupil of the eye, and he saw that the whole apparatus arose neither from the optic nerve, nor the retina, but from the purple, blood-rich layer – strangely reminiscent of the inside of a grape – which lies just behind the white of the eye.

Optics was not Felix's subject, but he was not without intelligence or imagination. It seemed clear enough to him what was going on: the crystalline humour 'is the looking glass of the optic nerve; and, placed before the nerve and the pupil, it collects the species passing into the eye as rays and, spreading them over the whole of the retiform nerve [retina], presents them enlarged in the manner of an interior looking glass, so that the nerve can more easily perceive them.'[7]

This is wrong. Platter forgot that magnifying glasses only work to magnify an image when used at close quarters. Hold a magnifying glass above this page and the words will appear larger, because light fans out from the page at all possible angles; the glass gathers all the rays that hit it, bending them back towards your eye. To the eye, it appears that light from the object is coming in from many more angles – which is to say, that the object is larger. However, rays of light hitting the lens from a distant object are, to all intents and purposes, running parallel to each other. The lens still gathers up this light and makes it converge at a focal point but because the rays are already running parallel, they come together much closer to the magnifying lens than would diverging rays. What the eye sees is an image produced behind the focal point – in other words, an inverted image.

Platter's anatomy is, none the less, very good, and it was only sensible that Kepler, encouraged by his old friend Johannes Jessen, should put his trust in his findings. Kepler's theory of vision applied the rules of optics to Platter's model of the eye. Light hitting the pupil is refracted, making the light rays that pass into the eye run more or less in parallel. When this parallel beam passes through the crystalline humour (the lens), it is once again refracted, and produces an inverted image upon the retina. From Platter and Jessen, Kepler was familiar with the idea that the retina is an extension of the brain, connected to it by the optic nerve, so he had no difficulty in asserting that it is the retina that receives images. But how such inverted images are perceived the right way up is a matter, not of optics, but of interpretation: a mapless and mysterious territory into which Kepler, wisely, did not venture, saying 'the armament of opticians does not take them beyond this first opaque wall encountered within the eye.'

So Kepler's innovation was, ironically enough, to apply already known facts to an already mapped structure – less of a discovery, more a brave insistence upon the inevitable. Still, we shouldn't underestimate the intellectual *chutzpah* involved. Kepler took considerable pains to put the retinal image the right way up again, but eventually abandoned the attempt in disgust. ('And there was no end of this useless labour', he complains.)[8]

We do not know to what extent Kepler was taxed with the implications of his own work. His focus was directed outward, at the motions of the planets. He drew no diagram to illustrate his theory of how we see, and the image traditionally used to illustrate Kepler's work is from the Enlightenment philosopher Rene Descartes' *La dioptrique*, of 1637.

It is a fitting and provoking choice. The optics are Kepler's and in all essentials they match our current understanding of how light behaves within the eye. But there is another detail, a flourish added by Descartes. At the bottom of the figure is an observer, studying the back of the eye, which has been scraped clear to make visible the inverted image within. On one level, this figure simply stands in for the anatomist. (Descartes himself once stripped the sclera and choroid from an ox's eye and used the eye to project an image on to a piece of paper.) At the same time, the figure also suggests a contemplative intelligence, standing somewhere behind the eye. This personification of the conscious mind cannot but remind us of the dreadful conundrum faced by all theorists of vision – including Descartes – who came after Kepler. If everything we see is a representation of the world, then seeing becomes just representation. If I cannot apprehend the world directly, but can only percieve representations of it, then where exactly am I, and where, exactly, is the world?

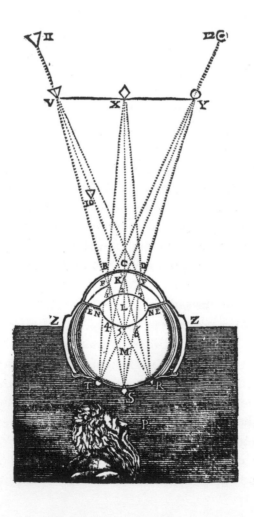

SEVEN

Nervous matter, visually endowed

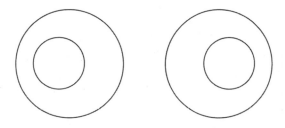

Bodies are not designed machines. They are palimpsests of half a billion years of gradual change. After this long, no anatomical structure performs just one function.

When the Roman anatomist Galen noted the mesh of major blood vessels lying across the retina, he concluded that the retina existed to feed the eye. For Galen, the retina *was* the mesh of blood vessels. That beneath these vessels lay another, nervous, layer only became apparent when Kepler's optical description of the eye insisted that the retina must be, in anatomist William Bowman's delightful phrase, 'a sheet of nervous matter visually endowed'.

The story of the discovery of the retina is vexed and broken. To tell it is to set many plates spinning at once. I hope that they continue to spin in your memory by the time I gather them up at the end of the chapter.

We begin with the story of 'a splendid toy', the ophthalmoscope, which did so much for our understanding of the retina, but did not let us see it. The second part describes how the elusive tissue of the retina finally gave itself up to close study, but told us virtually nothing. The extraordinarily beautiful drawings of the Spanish histologist Santiago Ramón y Cajal were required before anyone could begin to really understand how nerve tissues ticked. Meanwhile the infant science of physiology set about explaining the retina's working, and racked up some impressive successes. The final part of this chapter describes how the physiological observations of Ernst Mach, in 1867, were happily married to the anatomical work of Haldan Hartline, sixty-five years later. This marriage ushered in our current model of the retina.

1 – A splendid toy

Everything in the vertebrate eye means something.

Gordon Lynn Walls

Immediately behind the retina, cupping the pigment-stuffed tips of the photoreceptive rod and cone cells, lies a single layer of cells, the pigment epithelium. Its heavy black pigment, like the backing plate of a camera, stops light scattering off the back of the eye.

Nocturnal vertebrates do not have pigment epitheliums. They have mirrors. By bouncing light back through the eye this mirror, the tapetum, effectively doubles the amount of light the eye can see.

Strap a light to the barrel of your gun, and wait for night. Stand behind a likely tree, and aim your light-cum-rifle into the darkness. Wait for two bright discs to light up in the dark, and fire. The discs are the eyes of your prey: you cannot miss. This method of hunting, called pit-lamping, is an old prospector's trick. In the days before electric lamps, prospectors lit their way using lamps powered by acetylene and water. The lamp doubled as a handy hunting device, hence 'pit-lamping'. Strictly speaking, you don't need a lamp, or even a rifle, to go pit-lamping. All you need is light. When the native Americans of the Shuswap River in British Columbia go night-fishing for salmon, they light fires in their boats, using wood that doesn't crackle. Salmon eyes are good reflectors, and the fishermen spot and spear their prey by firelight, with very little trouble.

In the interests of a quiet life, I should point out to my Canadian readers that if you go out pit-lamping, you will be arrested. There are a number of reasons for the ban. There's the safety issue: a lamp's beam can pick out eyes at a great distance, but it isn't nearly powerful enough to reveal the whole animal. The eyes you're aiming at might be the eyes of a deer; but they might just as easily be the eyes of your neighbour's cow. But I suspect the real driving force behind the ban is the fact that pit-lamping is incredibly easy; game stocks would plummet if every fool with a lamp and a gun could bag his or her limit every season.

For most of human history, it was assumed that bright eyes were generating their own light. As late as the middle of the seventeenth century, Thomas Willis, the co-founder of the Royal Society, met an old man who claimed to read books by the light coming out of his eyes.

The problem was not that mirrors in the eye were unthinkable; animal mirrors – in the form of fish scales – are familiar enough, and anyone curious and strong-stomached enough to dissect, say, a horse's eye, will find its back wall covered with a shimmery coating. The problem – long before photoreceptors were discovered – was to understand how mirrors could possibly benefit vision.

What were seventeenth-century 'mechanical' philosophers and anatomists to make of mirrors in the eye? If the eye existed to gather light, why should it reflect all that carefully gathered light back into the environment? Even more puzzling, why should nocturnal animals, starved of light, waste this carefully gathered light so recklessly?

The difficulty of interpretation is evident in the *Anatomia Humani Corporis* (1685) by Professor Govert Bidloo. Bidloo (1649–1713), a playwright turned anatomist, served as personal physician to William III, first in Holland and later in England, where he was elected a Fellow of the Royal Society. His *Anatomia* is generally regarded as the finest anatomic work of the seventeenth century. In his discussion on the eye, Bidloo observes that the 'glowing' nocturnal eye is really emitting no more light than it receives. The careful wording is important: nowhere does he say that the eye is merely reflecting light. Bidloo kept his options open.

The Academy of Sciences, Paris's rival to the Royal Society, was founded in 1666 by France's finance minister, Jean-Baptiste Colbert. The academy immediately distinguished itself by the quality of its anatomical work. Anatomists like Claude Perrault and Jean Mery were at the forefront of efforts to advance anatomy as a comparative science – efforts which were greatly aided by the regular dissection of rare animals from King Louis XIV's private zoo.

In 1703, Mery noted that a cat's eyes shine much more brightly if you hold the cat under water. This – if you can get past the image of the experiment itself – suggests that eye shine has at least something to do with optics. But even here, at the birth of modern anatomy, Mery – a hard-nosed Mechanical if ever there was one, a man who undertook clandestine dissections whenever the opportunity arose – was not prepared to explain eye shine purely in terms of mirrors and optics, and regarded his findings as 'mysterious'. (The mystery is easily punctured. The submerged eye is not actually any brighter. It is simply that an eye in water no longer refracts the light to the same degree, so instead of being emitted in a beam, light sprays out of the cat's eye, making its shininess easier to spot.)

Karl Asmund Rudolphi (1771–1832), a Swedish-born naturalist, is best known for his *Grundriss der Physiologie*. This work, which argues that the human genus should be divided into species, not races, is notorious for lending a racist impulse to German and Scandinavian intellectual life, with dire and well-known consequences. In the year of its publication, 1821, well over a century after Mery dunked his cat, Rudolphi turned his attention to the directionality of the shining eye. He was able to show that the reflecting eye will emit light along exactly the same line as the direction of the ingoing rays. No chemical or biological process is taking place – a point he demonstrated by the simple expedient of shining lights into the eyes of a decapitated cat.

Rudolphi's macabre experiment brings the mystery of the reflecting eye to a head, as it were. Why should a light-starved eye reflect the light? Rudolphi made the matter no less mysterious: he did, however, suggest a way forward for research. The eye reflected light back along the line of illumination. If there were a way of capturing and studying this reflection, then it would be possible to study the back of the living eye.

Jan Purkinje – who just happened to be Rudophi's son-in-law – attempted this in 1823:

When I observed the eye of a little dog from a certain direction, that light seemed to be thrown back, until I discovered that the light is reflected from the hollow surface of the lens into the eye and then returned. When the experiment was immediately repeated with human beings, the same phenomenon occurred: indeed, the whole pupil lit up in a beautiful orange colour.[1]

What Purkinje is describing in such fascinated detail has become, in flash photography, a familiar and rather annoying problem: 'red-eye' – the ability of even the human eye to give off a bright orange-red glow if exposed to powerful light.

When he came to study the inside of the eye Purkinje, like many a researcher after him, kept getting in the way of his own light. How was an observer to place his line of sight along the same line as the line of illumination?

The physicist James Clerk Maxwell (1831–1879) seems to have solved the problem by 1854, but quite how is a mystery. In a letter to his aunt, Jane Cay, he wrote: 'People find no inconvenience in being examined, and I have got dogs to sit quite still and keep their eyes steady. Dogs' eyes are very beautiful behind, a copper-coloured ground, with glorious bright patches and networks of blue, yellow, and green, with blood-vessels great and small.'[2]

Maxwell did not pursue his enquiry, aware, perhaps, that several competing mechanisms were doing the rounds of the learned societies, all claiming to solve the illumination problem and thereby make it possible to study the inside of the living eye. Ironically, the best of these early 'ophthalmoscopes' predated Maxwell's doodling by about seven years, but no one knew about it.

Charles Babbage (1791–1871), computer pioneer and inventor of gigantic mechanical calculating engines, had a talent for poverty. Having invented everything worth inventing – from a speedometer, to a heliograph, to a cowcatcher – he expected at least a knighthood for his labours, but at the age of seventy-nine, impoverished, maddened in his last hours by the din of Italian organ-grinders, Babbage died, alone and uncelebrated. Among the inventions that ought to have made his fortune was a solution to the illumination problem: in 1847, curious about the double vision he experienced in one eye, Babbage invented the ophthalmoscope.

Babbage took a piece of mirror and scraped off two or three small spots of the silvering. He mounted the mirror in a brass tube, at an angle of forty-five degrees. An opening in the tube let light in. Light hitting the mirror was reflected down the tube into the patient's eye. Light reflected and focused by the eye travelled back up the tube to the centre of the mirror, passed through the clear spots in the glass, and arrived at the observer's eye. That was the theory, but to establish the practical usefulness of his invention, Babbage needed a professional champion.

William Mackenzie, the Glasgow doctor whose *Practical Treatise of the Diseases of the Eye*, of 1830, had become a standard work, exchanged letters with Babbage about his double vision, and recommended the eminent ophthalmologist Thomas Wharton Jones. Wharton Jones (who had earlier got himself unwittingly entangled in the Burke and Hare body-snatching scandal) was one of London's leading 'eye men'. Edward Jenner called him one of the greatest Englishmen who ever lived. There was only one problem: he was short-sighted.

When he looked down the barrel of Babbage's invention and saw nothing of what he had been promised – no network of blood vessels, no radial structure, no sign of a nerve head – Wharton Jones advised Babbage that his invention was useless. For this, Wharton Jones has come in for not a little retrospective stick. This is not entirely fair. Babbage had invented an ophthalmoscope; it just wasn't a very good one. The beam of light issuing from the eye is slightly convergent; had Babbage thought to add a weak diverging lens just behind the mirror, the images gathered by his instrument would have been a great deal easier to see: even Wharton Jones might have been impressed. As it was, the disappointed ophthalmologist saw nothing but a red splodge.[3]

He did, however, keep detailed notes of the demonstration. In 2003, these were followed, and a copy of Babbage's instrument was made for The British Optical Association Museum in London. It is clear from this that the first commercially successful ophthalmo-scopes resemble Babbage's prototype far more closely than did the instrument generally held to be the first ophthalmoscope – the one cobbled together four years later from cardboard, glue and glass slides from a microscope set by Hermann von Helmholtz.

Helmholtz (1821–1894), a physiologist with ambitions for a career in physics, arrived to run Königsberg's physiology department in 1848. Jan Purkinje's work on illuminating the eye had, by this time, largely been forgotten, but a student at Königsberg, William Cummings, independently stumbled upon many of the same findings. Helmholtz was intrigued by Cummings' work, and it was while he was trying to work out how to demonstrate it to his students that he threw together his 'splendid toy'.

Splendid it was: in its more finished incarnation, the ophthalmoscope revealed the back of the eye in exquisite detail, assisting greatly in the diagnosis and treatment of eye diseases. Its popularity was slow to build, however. Helmholtz enjoyed playing the showman among his acquaintances. ('The ophthalmoscope is, perhaps, the most popular of my scientific performances,' he once remarked.)[4] But Helmholtz, a physicist and physiologist, took little interest in the modifications and improvements necessary to make the instrument a commercial success. In the year following its manufacture, only eighteen ophthalmoscopes were sold.

What needed the most work was the light source. Having one's eyes examined by an ophthalmoscope powered by oil, gas, and even petrol must have daunted many a client. It says much for the constitution of the nineteenth-century public that the device eventually found favour as a parlour game.

2 – An invisible quarry

If you are going to study how the brain works, it is a good
idea to start with the simplest part.

                                        Jeremy Nathans

In the living eye, the retina is invisible.

Not only are its neural layers thin (wavelengths of light are thicker), they are transparent. More than that, they have optical properties virtually identical to those of the vitreous humour which fills the eyeball.

The match is so close, light passing from the humour into the nervous tissue does not glint or shimmer. (If it did, the consequences for vision would be disastrous.) However good the magnification, the study of the living retina reveals barely more than was visible to Galen, 1,800 years ago.

The opthalmoscope's contribution to understanding was considerable but, as we shall see in a moment, it was indirect. You cannot see the retina through an ophthalmoscope.

What can you see? You can see the larger branches of the blood vessels forming a mesh across the back of the eye – a spider's web centred on the head of the optic nerve. Because this is the retina's only other visible feature, it is very tempting to suppose that light is brought into focus here. Leonardo thought so, with good reason. If light is focused on the nerve head, then this can be the 'seeing' part of the eye, and the rest of the retina, with its mesh of blood vessels, can go on being a simple blood delivery system.

In 1619, the Jesuit astronomer Christoph Scheiner spotted the problem with this common-sense explanation: the optic nerve head can't be the 'site of foveation' because it is in the wrong place. In his drawings, Scheiner put the optic nerve head several millimetres towards the nose, away from where light is brought to a focus against the retina. It is the retina, and not the nerve head, that is 'seeing'.

Prince Charles Stuart, the second son of Charles I of England and Henrietta Maria of France, arrived in London to claim the throne on his thirtieth birthday, 29 May 1660. He received a warm welcome. England's experiment with republicanism had collapsed. But the clock could not be turned back. Charles II found his powers and privileges severely limited by Parliament, leaving him in the humiliating position of having to rely, for day-to-day funds, on a pension from Louis XIV of France.

Charles, made canny by his many years in exile, concealed his absolutist ambitions well. Only now and again did he behead his ladies-in-waiting. He learned the trick from his acquaintances at the Royal Society, to which he gave his royal seal of approval in 1661. They, in their turn, learned it from the French priest and physicist Edmé Mariotte.

THESE
WORDS WILL
VANISH

STARE AT THE CROSS

The blind spot. Cover your right eye and focus on the cross. Move
the book towards you, and the message to the left will disappear.

Mariotte, at first believing that the nerve head must be the seat
of vision, discovered that – visually at least – the nerve head did
precisely nothing. It was blind: each eye had a small blind area in
its field of view, corresponding to the location of the nerve head.
Mariotte had discovered the 'blind spot'.

Close your right eye, focus on the cross and move the book slowly
towards you. At a certain point (and with a little practice) the words
to the left of the cross will wink out of existence, revealing the white
paper beneath. Move the book closer, and the words will reappear.

Why – other than in experiments like these – don't we notice the
blind spot? As David Brewster once wrote, 'We should expect, whether
we use one or both eyes, to see a black or dark spot upon every
landscape within fifteen degrees of the point which most particularly
attracts our notice. The Divine Artificer, however, has not left his
work thus imperfect . . . the spot, in place of being black, has always
the same colour as the ground.'[5]

Repeat the experiment with a red piece of paper, and the words
will disappear to reveal red paper. Repeat the experiment with wallpaper, and the pattern of the wallpaper will be revealed. Your visual
system is 'filling in' the blind spot with the surrounding texture. A
miracle? Not really. Peripheral vision is very bad at detail, and because
the blind spot is in the periphery of vision, textures can be mapped
over the blind spot very crudely, with no loss of fidelity.

Charles II liked winking at girls. He had a reputation for that sort
of thing. But if he winked at you, there was always the possibility
that he was thinking how you would look without your head.

*   *   *

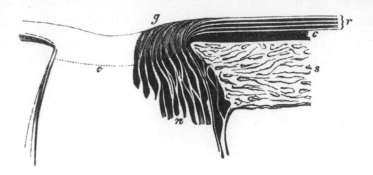

William Bowman's drawing of the optic nerve head. Bowman,
made a baronet by Queen Victoria for his services to medicine,
went on to found the Ophthalmological Society, which later
became the Royal College of Ophthalmologists.

The nerve head is blind: at this busy junction there is no room for
the apparatus of seeing.

So where exactly is this apparatus?

The Dutch microscopist Antoni van Leuenhoek hinted at the presence
of nerve fibres elsewhere in the retina, but the Italian naturalist and
physiologist Felice Fontana (1730–1805) is generally given the credit
for spotting them first. Fontana is best remembered for conducting
the first human brain stimulation experiments. By applying electricity
to specific regions of the brains of corpses, Fontana had been able
to make their faces spasm in interesting ways. When a law was passed
forbidding his work with the dead, Fontana advertised for – and got
– live volunteers. If Fontana had not existed, it would have been
necessary for Hammer Films to invent him: Fontana's love of anatomy
was such that he used to bring the bits he was studying to the dinner
table – a habit not without its risks, since among his discoveries were
the poison sacs of the viper.

As he was examining a rabbit's eye with a microscope, Fontana
spotted whitish bands running across the back of the eye, terminating
in a fibrous tangle at the nerve head. This was the first good anatomical evidence for the retina's visual function. Unfortunately, close
examination of other vertebrate eyes did not always throw up similar

findings. The human eye was particularly unenlightening: following Scheiner's meticulous drawings, we find that the image in a human eye is brought to focus not at the optic nerve, but at an area called the macula. The macula is a smooth depression in the retina, eminently suited to the role of seeing – except that there is no telltale white streak. At the centre of the macula, and barely the size of a pinhead, lies the fovea. But why should vision be concentrated upon such an absurdly tiny point, when the rest of the retina is clearly equipped for seeing?

It was the German naturalist, Gottfried Reinhold Treviranus (1776–1837) who, looking through a microscope in 1834, spotted a layer of densely packed cylindrical structures in the retina. Assuming (correctly) that these must be visual cells – photoreceptors, in today's terminology – he then had to explain how they functioned. The trouble was, in the heat of observation, he believed he had got his sample back to front. This seemed the only possible explanation for what he had seen: an array of highly specialised visual cells, covered with a mat of blood vessels. This arrangement was obviously nonsensical; Treviranus duly reported that his newly-observed photoreceptor cells must lie above the blood vessels, pointing up into the vitreous humour.

Treviranus had been right the first time. Three subsequent studies, by G G Valentin (1810–1883), J Henle (1809–1885), and E Brücke (1819–1892), showed that his cylindrical structures really did lie beneath a mat of blood vessels. These three then drew a conclusion as irresistible as it was wrong: since their view was obscured by blood vessels, these structures could not be visual cells, after all.[6]

Well into the nineteenth century, the best anatomists (and Treviranus was certainly one) were doomed to chase each others' tails, as they tried to unpick a structure too fine for their instruments to handle.

Strange then, that the experiment that finally settled the matter should be the ophthalmoscope – a machine through which one could not see the retina at all.

If by any chance you have an eye test looming, you may want to pay particular attention when your optometrist lights up the ophthalmoscope and brings it to bear on your eye. As light from the instrument enters your pupil at an oblique angle, the view of the darkened surgery will momentarily vanish in an orange-red haze, crazed by the tangled shadows of blood vessels. This is 'Purkinje's tree', an indirect glimpse of the eye's blood supply, which Jan Purkinje first conjured in 1823.

A glimpse of Purkinje's tree is as eloquent a demonstration of the vertebrate eye's oddity as you could wish for – what are blood vessels doing in front of the seeing parts of the eye?

A further mystery looms when the ophthalmoscope is moved about. As the light source moves, the shadows of the blood vessels shift with the change in the angle of the light. This means that the blood vessels lie some distance away from the seeing part of the eye. What fills this apparently empty space? One obvious candidate presents itself; the retina's elusive 'nervous matter', so easy to spot in the white 'visual streaks' of rabbits, but so oddly absent in the human eye. If the nervous matter connecting up the retina was excessively thin and optically neutral, then this would explain why it could not be spotted with microscopy, but only inferred from the waving branches of Purkinje's tree.

Purkinje's tree demonstrates that the seeing parts of the retina lie underneath two layers of tissue: a layer of blood-vessels, and a layer of invisible nerves. Strange as the idea seemed, it was more or less accepted by the time the German anatomist Heinrich Müller found a way to use fixatives to harden the retina. Studying the retina in cross-section, Müller found never-before-seen nerve layers between the retina's blood vessels and its photoreceptors. Unfortunately, only Müller could see them. Even his star pupil, Hermann von Helmholtz, could not be induced to see what Müller saw down the same microscope. For a while, these nerve layers were as controversial and contested as Percival Lowell's Martian canals, and for the same reason: because they lay on the very edge of visibility; fancy and expectation played as large a part in their mapping as actual observation. The retina's true complexity was revealed only in 1870, when Max Schultze (1825–1874), the German biologist and director of the Anatomical Institute at Bonn, made exquisitely detailed cross-sectional drawings. Schultze's retina, though literally paper-thin, boasted no fewer than ten neural layers.

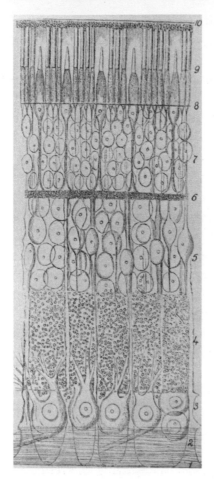

SCHULTZE RETINA
Max Schultze's 1872 drawing of
the layers of the retina.

Blood vessels lie over layers of nerve tissue, and layers of nerve tissue lie over the visual cells – this arrangement is common to all vertebrate eyes. The problem now was to explain how such a Heath Robinson arrangement could work.

Enter (yet again) Jan Purkinje.

An early riser, Purkinje would often wake before dawn and take a walk before breakfast. Everyone seems to have their own story about when and how he noticed the optical phenomenon known as the 'Purkinje shift'. Some say that, as dawn approached, he walked past his favourite geraniums and noticed with pleasure how the dark red blooms waved against a ground of pale green leaves. On his return he walked past the same flowers lit by the morning sun; now, pale red blooms stood out brightly against dark green foliage. Other accounts have it that the patterns of his bedroom carpet seemed altered at different times of day. The chances are that all these stories are true: Purkinje was a superb, eclectic observer.

Purkinje found that changes in light level affect how colours register on the eye. At dusk, blues, greens and yellows are brighter, while colours at the opposite end of the spectrum (reds and oranges) are dim. In bright daylight, the difference in brightness is levelled out and, in some cases, reversed, so that blues seem dim in comparison to reds. Purkinje drew the inescapable conclusion, published in 1825, that the eye contains two sets of visual equipment, one for daylight vision and one for dusk and dawn.

Where Purkinje's new-fangled experimental psychology led, anatomy eventually followed. In 1866, Max Schultze distinguished between two distinct types of visual cell in the retinas of birds. Because rod-shaped cells were more abundant in the eyes of nocturnal birds, Schultze proposed that rods were adapted for seeing in poor light. Birds that operated during the day had very few rods, but many more cones. So cones were the cells that saw in good light.

Considering the human retina, which contains healthy populations of both rods and cones, Schultze was able to improve upon Purkinje's explanation of his 'shift'. Rods enable vision at night; they are insensitive to colour, but respond in some degree to all visible wavelengths of light, giving us monochrome vision. Cones give us colour vision during the day, because each type of cone is sensitive to a different wavelength. If we average their sensitivities, we find that cones, as a class, react to longer wavelengths more strongly than rods do. During daylight, a red object appears lighter than a blue object emitting the same amount of light. At dusk, as we swap visual attention from cones to rods (a neat piece of visual orchestration and one that is still largely mysterious), the red object dims and the blue object appears brighter.

Because red is not a vivid colour in poor light, it is quite a bad choice of livery for emergency vehicles. Vehicles with green and yellow livery suffer fewer accidents. By the same token, red is an

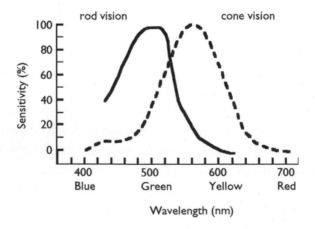

excellent colour for the instrument displays of ships and aircraft. In the dark, it is important that pilots can read accurately without losing their night vision. Since their rods are effectively blind to red light, pilots can read red-light instruments with their cones without blinding their night-adapted eyes.

From his work with birds' eyes, Schultze knew that cones and rods, in whatever proportion they happened to be present, were not mixed together willy-nilly. Cones tended to gather at the focus point of the retina, pushing the rods to the sides. (Some accounts say that there are no rods in the fovea. This isn't strictly true: only in the very centre of vision do cones push rods away entirely.)

If you are star-gazing, it is best to look very slightly to one side of the star you are studying. Your rods are much more sensitive than your cones, and will be able to pick up the object quite easily. Look at the star directly, and its image will fall on the dense nest of cone cells packing the fovea, so the chances are you won't see a thing.

Once we know that day- and night-time vision use different regions of the retina, a number of 'disadvantages' to the back-to-front retina fall away.

Yes, the rods are overlaid with blood vessels, and yes, the vessels certainly get in the way, but (saving ophthalmoscopic tricks) we can no more see our blood vessels than a camera, pressed up against a chainlink fence, registers the chainlink. To set against the slight blur that they add to an already blurred image, the blood vessels give the vertebrate eye the considerable advantage of an extra food supply.

Schultze noted that the optics of primate eyes perfectly focus just one degree of space. At this point, the retina is bent into a fovea – a shallow dish crammed with cones. Blood vessels, which overlie the retina everywhere else, do not intrude on the fovea. The area's relative thinness and lack of complexity – it is little more than a single layer of naked photoreceptors – means that it can be more than adequately fed from behind, by the choroid layer.

But the fovea's very simplicity is a source of mystery. Where are its neural layers? Why is the seat of focused vision also the thinnest, simplest part of the retina?

*   *   *

The Scots physicist James Clerk Maxwell probably contributed more to science than any other Victorian philosopher. Legends abound: it is said his work on the optics of index-graded lenses in fish was inspired by a breakfast of kippers. Occasionally scientific curiosity got ahead of him, as this passage, from a letter to his wife, Katherine, reveals:

> There is a tradition in Trinity that when I was here I discovered a method of throwing a cat so as not to light on its feet, and that I used to throw cats out of windows. I had to explain that the proper object of research was to find how quick the cat would turn round, and that the proper method was to let the cat drop on a table or bed from about two inches, and that even then the cat lights on her feet.[7]

If his list of key contributions omitted cats, it nevertheless covered fields as diverse as optics and liquid dynamics. He also made major contributions to our understanding of the fovea.

Maxwell was particularly interested in an unusual observation made by the inventor Samuel Soemmering in 1795. Soemmering noticed that the fovea and its immediate surrounds (an area called the macula) turned bright yellow when removed from the eye. At first, it was assumed that this startling colour change was caused by contact with the air. Maxwell wondered whether there was not another explanation. It could be that the structures and tissues that normally overlaid the macula protected it from a

James Clerk Maxwell.

particular wavelength of light. Once they were removed, this wave-length triggered the macula's change of colour.

Maxwell identified this wavelength by looking through a prism at a long vertical slit. As he tilted the prism, an elongated dark spot ran up and down the blue part of the spectrum, but could not be persuaded to pass through the other colours. The position of the dark spot indicated which wavelength of light was being blocked by the eye.

Maxwell's interest was further aroused by a report written by Wilhelm Karl von Haidinger (1795–1871), director of the Imperial Institute of Vienna. Haidinger described how, while studying a mineral sample through a polarising glass, he noticed a dark propeller shape hovering above the paper laid across his desk. When he rotated the filter above the paper, the propeller turned, too.

The figure set natural philosophers an irresistible challenge: another Scots physicist and inventor, Sir David Brewster and the mathematician George Stokes both essayed an explanation of 'Haidinger's brushes', as the propeller figure came to be called. But Maxwell had an advantage: he already knew that there was a structure in the eye that filtered out violet light.

To explain a puzzling visual phenomenon, Maxwell could usually be counted on to come up with a piece of apparatus, preferring demon-stration to argument. At a meeting of the British Association in 1850, he produced 'a small piece of apparatus which he had made by cementing together . . . sectors of sheet gutta-percha, these being so cut out of the sheet and put together that its fibrous structure . . . should radiate, at least approximately, from a central point. This arrangement . . . reproduced, or rather simulated, Haidinger's phenom-enon.'[8] His accompanying explanation went something like this:

Haidinger had been studying his sample through a polarising filter. That meant the light entering Haidinger's eye was polarised; all the light waves vibrated along the same plane. The dark brushes Haidinger noticed suggested that there was a second polarising filter, radial in structure, crossing his vision, and this could only be inside his eye, lying over the visual cells. When Haidinger rotated his polar-ising filter, he changed the angle at which the light in his eye was polarised, and the filter in his eye blocked the light along a different diameter; so it, too, appeared to rotate.

The subtlety of Haidinger's phenomenon held a further, final clue as to the nature of the filter in the eye. Haidinger's brushes did not block all the light. The brushes were not black; they were the faintest of shadows. So, the filter did not block all the light, just a portion. Maxwell, calling to mind his prism experiments, suggested, first, that the filter blocked only violet light, and second, that the active ingredient in the filter was macular pigment – the same pigment that Soemmering saw turn bright yellow when the macula was exposed to daylight.

Over the photoreceptors in the macula lay a single layer of incredibly fine fibres, radially arranged like the spokes of a wheel (or the gutta-percha fibres in his home-made apparatus). If these fibres contained macular pigment, then they presumably offered the seeing cells some modest protection against violet light.

So much for solving the mystery of Haidinger's brushes. The presence of radial fibres lying over the fovea was more suggestive still: it indicated how the fovea, so naked and unadorned, was attached to the rest of the retina. What if those very fine fibres were nerve fibres?

If they were, the fovea made considerably more sense. It was indeed attached to the rest of the retina, but as much overlying material as possible had been swept to the sides, so the cones could enjoy an unimpeded view. The fine, virtually invisible, fibres connecting the fovea to the rest of the retina also doubled as a filter, protecting the photoreceptors from ultraviolet light.

There is a discipline to good observation – one that fine artists spend their lives acquiring. Had the astronomer Percival Lowell been an artist, it is doubtful he would ever have persuaded himself into seeing canals. As it was, in 1895, his book *Mars* was published. 'Certainly,' Lowell wrote, 'what we see hints at the existence of beings who are in advance of, not behind us, in the journey of life.' His reputation never recovered.

In the same year a Spanish anatomist, Ramón y Cajal, on becoming a member of the Royal Academy of Sciences in Madrid, made an announcement that, for his own field, was just as shocking: the nervous system, he said, was made up of individual cells. Cajal (1852–1934) was born in the foothills of the Spanish Pyrenees. A vandal, hell-raiser and passionate artist, he would vanish into the hills for days with stolen brushes, paints and paper. With Cajal around, no

whitewashed wall was safe. His father Justo, a physician, was determined to stop his son frittering his life away in Bohemian poverty
and sent him to a school famous for its beatings and short rations.
To keep him out of trouble outside school hours, he was apprenticed
first to a barber and then to several shoemakers.

Cajal spent some years as an army physician. In Cuba, he contracted
malaria, then tuberculosis, and returned to Spain in poor health. His
father, by then Professor of Dissection at the University of Zaragosa,
got him work as a demonstrator of anatomy. But Cajal had never ceased
to be an artist. Now, at last, he found his proper subject matter: the
structures and tissues of the body. In 1877, 'using every peseta saved
from the service in Cuba', Cajal bought an old-fashioned microscope.

Why he chose the workings of individual tissues and cells for his
life's work is simple to explain. His health was poor, his finances meagre.
On these small structures, he could work cheaply, and from home.
Which, by all accounts, was no hardship. His marriage to Silvería Fañanás
García was a happy one: as well as his seven children, five of whom
survived into adulthood, we have the testimony of a friend who declared
'half of Cajal is his wife.' We can only regret that, despite several engaging
autobiographical writings – including a charming late memoir subtitled
'Reflections of an arteriosclerotic' – Cajal maintained the conventional
decencies of the period and said not a word about his family.

Cajal's interest in the nervous system was first provoked in 1887.
The psychiatrist Simarro Lacabra, knowing that Cajal was preparing
a book on laboratory technique, thought it worth showing him some
nerve tissue stained by a new method invented by the Italian neuro-
anatomist Camillo Golgi. Golgi's stain employed gold and silver
compounds that hardened within the tissue to be studied, leaving a
visible tracery. Golgi however, never quite mastered his own method:
his staining agent might stain one type of tissue in one sample, and
quite another type in another. Frustrated, Golgi nearly gave up the
use of his own invention.

Cajal turned the stain's fickleness to his advantage. So what if the
stuff stained one tissue, then another, then another? Was it not
obvious that, each time this happened, different kinds of tissue were
being stained? For Cajal, 'a look was enough . . . ideas boiled up and
jostled each other in my mind. A fever for publication devoured me.'

Before Cajal, most observers, faced with an undifferentiated mesh of uniform nerve fibres, had taken what they saw at face value and reported that the nervous system was simply a giant web or 'reticulum'. Cajal teased out the inter-relationships between nerve cells and traced the interweavings of neural communities. There is no question that he was vitally aided by Camillo Golgi's stain, but it was the quality of Cajal's observation itself – his artist's eye – that enabled him to lay out the nervous system, revealing that it was made up of many independent, but interlinked, cells.

The identification of many different kinds of nerve cell was provoking enough, but Cajal went further, showing how the nerve cells receive electrical impulses through incoming fibres, dendrites, and conduct signals to distant locations through outgoing fibres, axons. This meant that the nervous system was to be understood not just through the behaviours of cells, but through the relationships between cells of different kinds.

Golgi, a lifelong adherent to the idea of the reticulum, found Cajal's work outlandish. The two men met, for the only time, at Stockholm

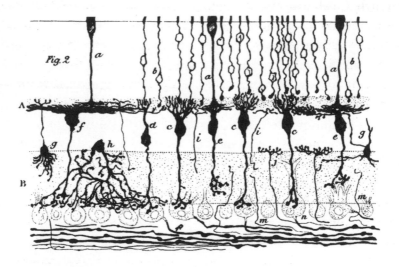

Cajal's drawing of a dog's retina shows rods (a) and cones (b) connecting to several different kinds of bipolar cell. Drawings like this put paid to the idea that the nervous system was a mesh of undifferentiated fibres.

in 1906, when they shared the Nobel Prize for Physiology or Medicine. This was the first time a Nobel prize had been shared, and the Caroline Institute's decision to split it between the two men in this way was almost scandalous.

'I thought that [Cajal] had deserved receiving a full, and undivided Nobel Prize,' Gustaf Retzius, a former member of the institute, remembered in his autobiography '. . . asked about this by the Nobel Council of the staff of professors at the Caroline Institute, I expressed this opinion of mine *decidedly*,' he writes, resorting to angry italics even after an interval of forty-two years.[9]

It is enlightening to compare the speeches of the two men. Golgi is clearly a product of the nineteenth century; whereas Cajal's speech might have been written yesterday. As the Swedish histologist Emil Holmgren reported to the Nobel Committee, 'Cajal has not served science by singular corrections of observations by others, or by adding here and there an important observation to our stock of knowledge, but it is he who has built almost the whole framework of our structure of thinking.'[10]

Though Cajal was always meticulously polite about Golgi and generous about his achievements, this was the civility we expect victorious youth to show to redundant old age. Golgi knew it, and hated it. When they received their award, Golgi spoke first, and used the occasion to defend the reticulum while at the same time stealing as much of Cajal's glory as possible. He convinced no one.

3 – Contrast

Every light is a shade, compared to the higher lights, till you come to the sun; and every shade is a light, compared to the deeper shades, till you come to the night.

John Ruskin

Ernst Mach (1838–1916) called himself a physicist. In 1864 he took a job as professor of mathematics in Graz, and in 1866 he was also appointed professor of physics. He is best remembered for working out what happens when the speed of a travelling object exceeds the

speed of sound, and the speeds of supersonic aircraft today are commonly given in multiples of the speed of sound: 'Mach numbers'.

Mach's early years in Vienna in the 1850s were dogged by lack of funds. He was hardly poor, but he could not afford to fund purely physical researches. Less well explored – and for that reason, more amenable to cheap and cheerful experiment – were those still-unlabelled areas where physics, physiology, and the psychology of sensations overlapped. In 1867, a professor and newly married, Mach made a series of simple observations, uncovered an underlying mechanism of vision, and transformed our understanding of vision, perception and the nervous system.

His breakthrough was to explain an optical illusion generated by a series of grey bands, each one a little lighter than its left-hand neighbour. (Mach used black-and-white spinning tops to create very precise shades of grey, but his findings are now more usually presented in printed form.) Each 'Mach band' is evenly printed, without grain or shading, and yet each band appears fluted. It is as though the bands have been lit from the side; the edges lying against darker neighbours appear lighter, while edges lying against lighter neighbours appear darker, as though in shadow. The fluting is an illusion – but why should the eye manufacture dark where there is no dark, and light where there is no light?

Mach realised that the eye is interested in boundaries. By exaggerating the contrast between neighbouring bars, the eye reveals the line along which they join. Mach worked out a detailed mathematical model of this process, suggesting how neighbouring points on the retina would interact with each other to create the Mach band illusion.

Of course, it was only a model; there was no guarantee that real structures in the retina behaved that way.

The American physiologist Haldan Keffer Hartline (1903–1983) received his medical degree in 1927, on the solemn understanding that he should never attempt to treat patients.

This was Hartline's version, anyway, and it is typical of the man's dry humour and eccentric reputation that he was widely believed. His laboratory was once described by his long-term collaborator, Floyd Ratliff, as a 'slightly disorganized but extremely fertile chaos',[11] and if he wasn't to be found there, then he was most likely engaged in outdoor adventures of one sort or another: though slightly built, Hartline was an athlete, with several first ascents in the Wyoming Rockies to his name. He also enjoyed sailing, and flew his own open-cockpit plane.

Though he once shared lab space with George Wald, the man who discovered how the retina used vitamin A, and sailed with Roger Granit, who found out much about the retina's electrical behaviour, the three men never collaborated. It took the Royal Caroline Institute in Stockholm to bring them together, when it awarded them the Nobel Prize in 1967.

In the 1920s, when Hartline began his studies, the discipline of anatomy was at an impasse. On the one hand, new tools were being developed to explore the workings of the nervous system – the last great anatomical unknown. 'Listening in' to the electrical activity of the nervous system was, according to the British medical pioneer Edgar Douglas Adrian (Lord Adrian of Cambridge), 'a new, very powerful microscope to work with'. On the other hand, if the things under study are inadequately prepared, nothing useful will be revealed, however good the microscope. In 1927, Adrian and his co-worker Rachel Matthews had managed to record the electrical activity of the optic nerve of a conger eel. This was an astonishing technical achievement, but the recording was meaningless, a white noise made up of the chatter of hundreds of thousands of individual fibres.

It had been known since 1865 that the retina reacted electrically to light, but to really unpick how the retina behaved would take recordings of single nerve fibres – and how could one possibly isolate a single fibre?

Hartline and his collaborator Clarence Graham claimed that it was by pure good fortune that they chose the horseshoe crab for their research. If true, this was definitely a case of chance favouring the prepared mind: it would not have escaped their notice that the horseshoe crab was ancient, a 'living fossil' that had been swimming, crawling and spawning in the Earth's tidal shallows for at least 250 million years. The closest living relative to the crystal-eyed trilobites, horseshoe crabs boast coarsely-faceted compound eyes, whose photoreceptors are a hundred times the size of human rods and cones (the biggest of any known animal), and optic nerves that can be frayed into thin bundles which are easy to split down to a single active fibre.

In 1932, Hartline and Graham managed to record the activity of individual fibres. They found that every nerve signal was identical. This meant that any information the nerve contained was being conveyed purely by its level of activity; the shape or amplitude of each signal was irrelevant. However, other, more anomalous results, suggested that something peculiar was happening to information at the level of the retina.

When Hartline began work on the horseshoe crab's compound eye, he assumed that each individual ommatidium, would work independently. But he soon noticed that extraneous lights in the laboratory, rather than increasing the rate of fire of a receptor, often caused it to fall silent. Why should light stop a receptor from firing? Hartline realised that the ommatidia were not independent. Each acted in concert with its neighbours. If one ommatidium was brightly lit while its neighbour was dimly lit, the dimmer signal became even weaker than normal. The result was a greater difference between the two signals – in other words, an enhancement of contrast.

Contrast seemed to be the only fact about the illuminant which found itself translated into electrical signals. Turn the lights up, and

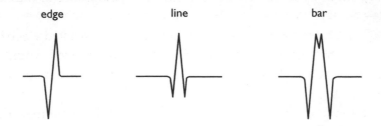

edge   line   bar

Mutual inhibition is evident in these sketched responses of
fields of ganglion cells, confronted with different kinds of edge.

the nerve fibres grew excited – for a moment – before settling back
to their rest state. Turn the lights down and the same thing happened.
At no point did the nerve fibres seem to care about the actual level
of illumination. Hartline was impressed: this 'contrast only' form of
vision meant that the eye could detect small local variations in light
intensity, even though ambient light varies a millionfold between
sunlight and starlight.

Mach's conjecture had turned out to be correct: contrast is accen-
tuated in the eye by a process Hartline dubbed 'mutual inhibition'.

Imagine a group of six photoreceptors, all attached to a single
nerve cell. It is dark. The photoreceptors are silent and so is the
nerve. The light goes on. All six receptors fire. For a moment, there
is a lot of activity, until the nerve cell issues a six-fold inhibiting
signal to its pool of receptors, and everything falls silent again.

Now, imagine that light catches just one receptor. The receptor
fires, but the other five, still in darkness, stay silent. The nerve cell
responds by sending a single inhibiting signal, *but this one signal is
shared between all six receptors*. This means that the lit photoreceptor
will remain fairly active while its quiet neighbours are effectively
damped. An illumined receptor whose neighbour is in the dark will
give a stronger signal than a receptor surrounded by other illuminated
receptors and a poorly illuminated receptor will have its already
small signal suppressed to virtually nothing if it lies next to a well-
illumined cell.

Now imagine a million nerve cells, each one overlapping its neighbour, each performing the same trick, making the dark side of an

edge darker and the light side of the an edge lighter – and you will
start to get some idea of how cells that inhibit each other actually
make vision possible.

Computer modelling can give us a good visual idea of what this
system achieves, as David Marr found during his ground-breaking

(a)   (b)

How the retina sees: David Marr filtered these images
using responses typical of nerve cells in the retina. Areas
of detail are revealed, while patches of constant illumination
are rendered an even grey.

work at the Massachussetts Institute of Technology. These images, from *Vision*, his celebrated account of that research, shows what happens to a black-and-white photograph when it is filtered using the kinds of responses typical of a particular kind of retinal nerve cell, called a ganglion cell. Large expanses of light and dark in the original photographs (areas which tend to reflect absolute light levels, but little else) are effectively evened out to a medium grey while areas of high contrast are massively exaggerated, even to there being flanking strips of white to the black wires of the chainlink fence.

Rendering a nice photograph down to something that looks suspiciously like a poor photocopy does not look like much of a trick. We should remember, however, that the eye has evolved to report salient features, not pretty pictures. By reporting only the lines of contrast, the retina avoids having to prepare endless, uninteresting, and massively redundant reports about plain surfaces.

We receive only local contrast information from the eye, and infer shading at our leisure. This point was demonstrated quite spectacularly in 2003, by Dr Steven Dakin of University College, London, when he filtered the iconic image of Cuban revolutionary Che Guevara to remove everything but the contrast information.[12] Look very closely at this image, and you will see that the lit parts of Che's face are exactly the same shade of grey as his beard and facial shadows. The lines alone reveal which side of each edge is supposed

to be bright, and which is supposed to be dark. Whether we look at the original black and white image, or the filtered image, we are using contrast cues to recover patterns of light and shade. In the discrimination of a scene, contrast is everything.

Enthused by his findings, Hartline turned from the eyes of horseshoe crabs to the eyes of vertebrates. The vertebrate eye is much more complex and compact, and the technology needed to tease out and listen to a single fibre from a vertebrate optic nerve did not yet exist. But Hartline realised that this was not a problem. The optic nerve was a conduit; all the really interesting activity took place at the retina, where the optic nerve was already spread out in a thin layer. The retina was a ready-made dissection of the optic nerve.

Through a combination of canniness and relentless practice, Hartline eventually teased out single nerve fibres from a frog's retina. He had expected the nerve cells to mutually inhibit, much like the nerve cells in the retina of a horseshoe crab. The reality was much more exciting: the frog's nerve cells reacted in many different ways to the light. Along with cells that reacted to contrast, there were cells that responded to changes in ambient light intensity, regardless of whether the light grew brighter or darker. Others reacted only when the light was dimmed; still others responded to the movement of dark spots or shadows. As the list grew longer, it became apparent that all these cells were, at heart, operating like the cells in the horseshoe crab retina. The difference was that the horseshoe crab processed its visual information once, across a single layer of nerve cells. The frog, on the other hand, boasted several layers of nerve cells, and visual information, analysed several times by successive layers, was processed to reveal more and more about the world: where objects were, which way they were moving, maybe even what colour they were. 'Individual nerve cells never act independently,' Hartline explained, during one lecture; 'it is the integrated action of all the units of the visual system that gives rise to vision.'

Before Hartline, it was usually assumed that the optic nerve carried just a single sort of visual information – a picture, if you will. Now we realise that even before visual information leaves the retina, it is being integrated and analysed for significant features. Different kinds of information are processed by different kinds of cell, and carried down the optic nerve along distinct pathways to distinct areas of the brain.

World War Two interrupted Hartline's work, and its value was not truly recognised until the early fifties. Since then huge strides have been made: cells in the frog retina were discovered which were capable of ever more complex responses; other animals were studied, and people began to appreciate how specialised each animal's retina was; cells which appeared identical under the microscope served completely different functions in different retinas, depending on how they were strung together.

Why does visual information need to be manipulated so soon – almost the moment light touches the eye? Why can't it simply be piped down the optic nerve to the brain? A quick look at the numbers reveals the answer. There are 126 million photoreceptors in the human retina, but only a million or so fibres in the optic nerve. Were every photoreceptor connected to the brain by a nerve fibre, the optic nerve would be as thick as the eyeball. This would be an horrendous logistical problem for any species, and particularly ruinous for humans, since it would make it impossible for us to move our eyes.

The more salient the information coming from the retina, the less of it we need. One million well-edited signals are better than 126 million chaotic ones. How the editing is done depends on the time of day. The retina contains two seperate mechanisms of vision (one might almost say that it is two retinas in one). One is for night (where sensitivity to any and all light is essential) and one is for day (where light of different wavelengths can be processed to generate colour vision).

At night, visual information is gathered from the rod receptors, which lie across the whole field of vision. Although there are one hundred and twenty million rods in the eye – rods are twenty times more numerous than cones – they do not generate twenty times the traffic down the optic nerve. To efficiently detect very faint levels of illumination, the rods are connected by nerve fibres into fewer, larger units. By pooling signals, the rods effectively become fewer in number and larger in extent. With so many rods merging their information before piping it down a single fibre of the optic nerve, there is a good likelihood that even small amounts of light will be gathered and registered. The price paid for this sensitivity is fuzziness: fields of rods are much larger than single photoreceptors and will inevitably produce an image with a coarse grain.

This balance between sensitivity and acuity is finely managed. Closer to the fovea, rod fields are smaller, as they attempt to catch fine detail; towards the periphery, the fields are larger, to detect the very faintest objects and movements. What is more, the arrangement is dynamic; as light levels fluctuate, the fields change size: at the threshold of human vision, pools of up to a thousand rods enable us to spot sources so dim and distant, it is said we can see the light of a single candle shining seventeen miles away.[13]

Daylight vision, on the other hand, prioritises information from the centre of vision, at the expense of peripheral vision. Rather than condensing information from the fovea, nerve cells may carry information from single cone cells straight to the optic nerve. (The 'grain' of the field of view is at its finest here.) It is usual for a cone to be connected to more than than one set of nerve cells; information from the cones is duplicated, and the fovea, which occupies a mere half of one per cent of the retina's surface, produces information so important that it takes up forty per cent of that part of the brain which first receives visual information.

Hartline and his successors produced an intricate picture of how vision works but it was a monochrome picture, a picture suited to night-time, and deep water, and eyes that struggle to distinguish form within the murk. Eyes adapted to take advantage of bright light boast complexities Hartline could not possibly have hoped to unpick.

'Avoid vertebrates because they are too complicated,' Hartline advised his students, 'avoid color vision because it is much too complicated, and avoid the combination because it is impossible.'

EIGHT

Seeing colours

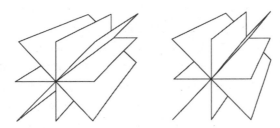

Whenever I mention that I'm writing a book about eyesight, I am invariably treated to the same story: how one day, as a child, it dawned on the person I am speaking to that their experience of colour might be entirely their own. Who is to say that my experience of green is the same as yours?

The near-infinity of the colour space we perceive makes colour a mysterious and personal sensation. We don't all see the same colours, any more than we all have identical tooth-aches, but just as we can agree on what a toothache is before comparing dentist stories, so we can all agree on what 'orange' and 'purple' and 'green' are before we go on to argue about the precise hue of the plaid our friend is wearing. We never agree entirely, but we always agree enough.

We can all match objects of the same colour and some are better at this than others. Many chemists can discriminate between colours as finely as wine connoisseurs discriminate between bottles of Pauillac. For now, let us agree that there is sufficient orderliness to colour for us to group objects together in consistent ways, and agree with each others' choices most of the time. Where we disagree, the disagreements are not crazy or puzzling. A colour-blind man (for reasons of heredity, it is usually a man) may put green and red objects together, but that's about as wild as it gets. No one has ever come up with unique multi-coloured categories. We map colour space in slightly different ways, but we are all mapping the *same* colour space.

Where does this orderly colour space reside? Are objects coloured? Does light come in different colours? Do we make up colours in our heads? Is colour a mere linguistic convention?

The first part of this chapter attempts to answer these and similar questions. The rest of the chapter is an historical account of colour vision. There is certainly a conventional way of telling that history but that may be no bad thing according to John Mollon, professor of visual neuroscience at Cambridge:

> Each newcomer to the mysteries of colour science must pass through a series of conceptual insights. In this, he or she reca-pitulates the history of the subject. For the history of colour science is as much the history of misconception and insight as it is of experimental refinement.[1]

My potted version of that history ends in 1973, with Edwin Land's 'retinex' theory of colour vision; then, in the final section, we leap-frog back to the workshops and studios of London and Paris in the 1880s, for it was there, among the artists of the Divisionist or *pointilliste* school, that experiments in colour perception laid the groundwork for the most fashionable field of current vision research – the study of visual attention.

1 – Colours and words

> I wish you . . . could eradicate the insane trick of reasoning about colours as identified by their names. People seem to think that blue is blue, and one blue as good as another.
>   C J Monro to James Clerk Maxwell, 3 March 1871

What colour are the walls in your kitchen? Are they nol? Or are they wor? To help you, *nol* is the word the Burinmo hunter-gatherer people of Papua New Guinea give to a sort of greenish-bluish-purple and *wor* is yellow. Well, yellowish-orange, with hints of brown and green. Which, funnily enough, is as good a description of the walls in our kitchen as any in the paint catalogue. (We got rather over-excited with our colour choices when we moved in.)

Are our kitchen walls wor? I'm not sure. Even if you showed me something certifiably wor-coloured, I might not be able to pick out the colour again. English lacks a name for this colour and with no easy label, it's devilishly difficult for me identify it. This was the finding of Jules Davidoff, a psychologist at Goldsmiths College, London, who is studying how language affects our perception of colour.

The Burinmo language distinguishes between nol and wor, but makes no distinction between blue and green. Shown a green and asked to memorise it, the hunter-gatherers found it difficult, after an interval of only a few seconds, to select it from a tray of other blues and greens. It's not that we don't see the colours we have no names for; it's simply that the colours we can name are lodged in our memory in a way that others are not. Nameable colours are the beacons by which we navigate colour space. The more names we learn, the more ably we navigate. This was brought home to me, with some force, as I researched this chapter. I have never, until recently, been able to identify indigo. Although I'm not alone in this – about half the people I spoke to about it share my incapacity – this failing has always niggled me. There are, according to no less an authority than Isaac Newton, seven colours in the rainbow, and I can only see six. At the blue-green end of the spectrum, there is supposed to be a band of indigo between the blue and the violet – so why has it been so hard for me to spot?

Indigo is an example of a colour word that has fallen out of fashion. It's in the dictionary, but hardly ever crops up in conversation. Had you pointed at something indigo, until last week I'd have seen a dark blue. Questioned further, I would have happily conceded that there was an element of red in the blue, but didn't that make indigo just a fancy name for purple? Not any longer. Research for this chapter has piled my desk with any number of examples of different colours, arranged in any number of colour systems. I'm surrounded by a veritable swatch table of indigos. And because they're all helpfully labelled 'indigo', the colour has begun to lodge in my memory. Last week I

walked into a Sainsbury's supermarket, and what did I see but Sainsbury's store livery: orange on – miracle! – indigo.

So how many colours are there in a rainbow? It depends to whom
you speak: Newton, it must be said, had a vested alchemical interest
in there being seven distinct colours. Though I recognise indigo, I
can't honestly say that it makes a distinct band between blue and
violet. Similarly, the orange and red bands seem to me to be so
hopelessly merged that there is no separating them.

The Greek poet Homer praises:

> Jove's wondrous bow, of three celestial dyes,
> Placed as a sign to man amid the skies

The philosopher Xenophanes (c.570–465 BC) mentions three colours:
purple, yellow and crimson. I don't know how many colours the
Burinmo people see in a rainbow; not many, I'd wager. They only
have five colour words and two of them are nol and wor.

It seems that, the more insulated the society, the fewer colour
words it will employ. In 1969, the anthropologists Brent Berlin and
Paul Kay wrote a description of how colour terms arise. The most
isolated societies do little more than distinguish light from dark. As
societies emerge into the wider world, red is the first 'true' colour
they identify, followed by a greenish-yellow. Green and yellow are
invariably distinguished before blue is distinguished from black. The
rest follow in no particular order. We don't have to go anywhere very
exotic to see elements of this scheme in operation, for most modern
languages contain traces of the evolution of colour. For example, up
until the Industrial Revolution, the Welsh 'glas' meant 'mountain-
lake-coloured', and did for both green and blue. Over time, the influx
of new dyes and textiles, English influence, and growing urbanisation,
made it expedient for the Welsh to distinguish between green and
blue. In modern Welsh, 'glas' simply means blue; 'gwyrdd' means
green. The native Japanese language has just four native colour terms;
only late in its history did it start borrowing others from Chinese
and English. The trace of these borrowings is evident, since the loan
words obey different grammatical rules.

If Berlin and Kay's scheme has fallen out of favour with today's anthropologists, it is not because it is 'wrong' exactly; it is rather that it has proved insufficient. Many languages don't distinguish colours from other visual properties: texture, reflectance and brightness are often described using terms that double as the names of colours. The translator's job is made even harder by the various vicissitudes and accidents to which words are prone. Often, colour terms derive from the names of dyes and dyestuffs. (Just a couple of hundred years ago, 'pink' was a noun.) And since a dyestuff often generates more than one colour, this may be why the English word 'blue' derives from 'flavius', a Latin word meaning yellow.

Have special pity for anyone translating an ancient Greek text. So-called Greek 'colour words' have no direct English equivalents. Worse, they don't refer to colours, relating more to texture, consistency and quality, with colour a small, often irrelevant, part of the whole meaning. The sea is the colour of wine, but so are sheep. Honey, sap and blood are all *chloros* which, as far as we can tell, is a sort of yellow-green.

But what green recruit has not felt blue at times? The English language is not short of colour metaphors, and neither was ancient Greek. If we look for a wider meaning for 'chloros' we find that it can also mean 'fresh', 'fluid', and 'living'. Studying these metaphorical meanings, we begin to see what the devil the poet is on about.

The odd thing about Greek poetry (and Homer's *Iliad* in particular, the best, earliest and most substantial example) is that the metaphorical meanings are the only meanings employed. Why does Homer never speak directly and simply about colour? William Ewart Gladstone (1809–1898), four times British Prime Minister under Queen Victoria, and a great classicist, was unequivocal in his criticism of Homer's colour palette:'Although this writer has used light in various forms for his purposes with perhaps greater splendour and effect than any other poet, yet the colour adjectives and colour descriptions of the poems are not only imperfect but highly ambiguous and confused . . . we find that his sense of colour was not only narrow, but also vague, and wanting in description.'[2]

From this, Gladstone leapt to some rather sweeping conclusions. Consider his book, *Homer and Homeric Age* (1858), the first sally in a life-long attempt to reconcile the literature of heroic Greece with the literature of the Bible. Not content merely to remark upon the green sheep and wine-dark seas of the *Iliad*, Gladstone – with youthful enthusiasm – leapt from the particular to the universal with two extraordinary claims, one far exceeding the other for sheer *chutzpah*. First, he said that the ancient Greeks were colour-blind. This is by no means a silly idea. Colour-blindness is a possible explanation of Homer's choice of colour terms, for reasons explored in the next chapter. Gladstone's second claim, however, far outstrips the first: that among the Greeks '. . . the organ of colour was but partially developed.' He believed he had discovered *prima facie* historical evidence of the way human vision had evolved. According to Gladstone, what we think of as normal colour vision arose after the composition of the *Iliad*.

The enormity of the idea was irresistible: here, at last, was cultural evidence of human evolution. The German philologist Lazarus Geiger was inspired by Gladstone's claim to investigate Greek literature, the Vedic hymns and the Zend-Avesta, amongst other writings, to see whether the development of colour vision could be traced in other ancient literatures. Needless to say, he found what he was looking for:

> All, at first, was vague in color . . . but gradually a difference was perceived, and men were compelled to find some term to express this newly observed appearance . . . green was for a long time regarded as yellow . . . Not only was the sky not called blue, but nothing was called blue, and it was impossible to call anything blue . . . the men of that time did not and could not call anything blue.[3]

As the years passed, scepticism grew. Advances in archeology left Gladstone's attempts to reconcile mythology with the Bible looking rather risible. (In her novel *Middlemarch* (1871–2), George Eliot neatly captures the contemporary mood with her portrayal of the pedant Casaubon, who has devoted his life – and, by-the-by, his young wife's – to 'The Key to All Mythologies').

And aside from the historical niceties, there is always the possibility that Homer didn't want to talk about colour. The idea of pure colour could hardly have existed in the ancient Greek world. Only a handful of pigments were available to them: white, blue-black, red and yellow-green. These, according to Empedocles, were the primary colours: those from which all others are made. (If you play with the precise pigments to which he refers, you really can generate a full spectrum of colour, albeit a rather muted one.) The Greek world was not saturated, as ours is by artificial colour. The *Iliad* is a rich compendium of surfaces, reflections, mists and tricks of the light. Why should the ancient Greeks have separated out colour for special emphasis, rather than texture or lustre?

Perhaps the Greeks – or Greek writers – did not consider colour very important. Seven hundred years after the composition of the *Iliad*, in the third century AD, Heliodoros managed to write a sixty-thousand word romance, the *Aethiopica*, without once using the words red, green or blue. This same lack of interest has been encountered recently; in 1971 a team of Danish anthropologists went to Polynesia to study colour perception among the islanders. But in one village, they were told, 'We don't talk much about colour here.'[4]

In 1996, I drove north from Helsinki towards the Arctic Circle. The Finnish countryside is a curious, fractal landscape of lakes and forests, repeating endlessly, with minimal variation. Either the tedium drives you mad or you achieve a kind of *satori*. It was in this Zen mood that I reached Rovaniemi and the borders of the Arctic. I parked the car, wandered around Santa's house for a little while (why come all this way and not say hello?), bought an outrageous hat, and walked back to my car. I had still to find an hotel before I began the final, surreal, reindeer-dodging part of my journey through the tundra to the rock-strewn coast around Lake Inari. It was evening. I stepped out from behind the shadow of Santa's grotto and stopped, pole-axed.

The sun was setting. It was October, when the days are more or less the same length they are in Britain. It was the wrong time of year for the midnight sun, the Northern Lights, or the other heavenly exotica for which these latitudes are rightly famed. What brought me to such an abrupt halt was an ordinary evening and an average sunset – except the colours were all wrong. The sky was no longer the sky I recognised. It was neon. The pink edging of the clouds was an artificial hot pink. Where I expected orange, there were flushes of peach and cherry-red. The sky itself had a slight violet cast. It was a sky painted by a painter who, to my eyes, had never seen a sunset and was relying on written accounts. It was a sky painted, perhaps, by Akseli Gallen-Kallela, the master of Finnish landscape painting, whose early, bucolic works had left me cold and unconvinced when I had viewed them, days before, in Helsinki's Ateneum Art Museum – and now I knew why. Gallen-Kallela's strange, forced, artificial choices of colour were the right colours for the Finnish landscape. This far north, things are lit differently.

The Finnish sky is more blue than the southern English sky. The more atmosphere sunlight passes through, the more its shorter, blue wavelengths are scattered by air molecules. (If this didn't happen, the daylit sky would be as black as the interior of Arthur Zajonc's light box.) The sun's rays pass obliquely through the atmosphere to Finland, making its blue sky particularly intense; to reach southern England, sunlight follows a more direct path. The sunlight's shortest route through the atmosphere is at the equator, where the skies are more white than blue. (The noon sky – when the sun is directly overhead – is less blue than in the morning or late afternoon.)

When we look at the sun at dawn or sunset, it has a distinct reddish-orange cast, because all the short, bluish wavelengths have been scattered, and only the longer wavelengths are reaching us along a direct path. The contrast between the reddish sun and the blue sky makes for pleasing sunsets in England, rather dull fare at the equator (the sky is at its least blue here and the sun at its least red), and firestorms of cherry reds and hot pinks in Finland.

How many colours are there? In Vincente Minnelli's 1956 biopic of Vincent Van Gogh, *Lust for Life*, the tortured artist chases across Europe after an elusive 'light', for the very good reason that this is exactly what artists of that time did. J M W Turner's obsessive pursuit of different lights was so well-known it became the subject of newspaper parody. Nineteenth-century Europe saw an explosion of novel colours as synthetic pigments immensely extended the artist's range of possible effects. Industry was colouring the world in inadvertent ways, too. I live in Sydenham, a humble London suburb immortalised in 1871 by the Impressionist painter, Camille Pisarro. Pisarro, like

Claude Monet, came to London for the pollution. The smoke belching from the capital's chimneys was an irresistible draw, generating colours and light effects that were not just odd, dramatic, and atmospheric, but new.

That there might be physiological limits to the number of possible colours is a fact not at all apparent to the eye, which seems capable of discriminating between an infinite range of hues. Is there, somewhere in the world, a never-before-seen colour lurking on the wings of a tropical bird, glinting in the scales of a reef-dwelling fish, or buried deep in the lustre of a rare gem fresh from the Mexican mine? How can we be sure that there are no more new colours to find?

2 – Primary colours

Seventeenth-century philosophers were interested in the behaviour of light, and the coloured lights visible through prisms were a particular source of fascination. But if science was going to say anything useful about light, it had to decide what it was competent to study. How, in the seventeenth century, was one to hunt down the mysteries of colour, let alone measure them?

Newton's Dutch contemporary, Christiaan Huygens (1629–1695), clearly delimited his philosophy's powerful but narrow abilities: 'In true philosophy,' he wrote, 'one conceives the causes of all natural effects in terms of mechanical motions.' The trick, for seventeenth-century optics, was to separate what was quantifiable – the mechanical behaviour of light – from the ineffable experience of colour in the eye and heart of man. This was not a denial of the subjective realm, but a recognition of what the science of the day could and could not do.

In 1666, in a room of his birthplace, the manor house of Woolsthorpe, Lincolnshire, Isaac Newton (1643–1727) began experimenting with the 'celebrated phenomenon of colours'. Newton knew full well that he would not, by the light of the prevailing philosophy, be able to say much that was useful about the phenomenon of colour. 'Rays, to speak properly, have no Colour,' he wrote. 'In them, there is nothing else than a certain power and disposition to stir up a sensation of this Colour or that.' This intuition – that colour is not a property of things but is generated by the eye itself – is spot on; and the slightly desperate hand-waving ('a certain power and disposition') marks the point at which the niceties of objective measurement leave off and the woolly posturings of medievalism take over: posturings he was determined to avoid.

Newton's experiments with prisms – a common toy and fairground gewgaw – were directed not at the colours, slippery beasts that they were, but at the shape a beam of light made as the prism fanned it out. 'It was at first a pleasing divertissement to view the vivid and intense colours,' Newton wrote to Henry Oldenburg, Secretary of the Royal Society, 'but after a while applying myself to consider them more circumspectly, I became surprised to see them in an oblong form; which, according to the laws of refraction, I expected should have been circular . . .' A neat, round beam of light enters a darkened chamber through an aperture. It passes through a prism, and makes an *oblong* pattern on the screen beyond. 'Comparing the length of this coloured spectrum with its breadth, I found it about five times greater, a disproportion so extravagant that it excited me to a more than ordinary curiosity of examining from whence it might proceed.'[5]

To better understand what the prism does to the light, Newton drilled a narrow aperture through the screen, and allowed a narrow portion of the oblong to pass into a second darkened chamber, and through a second prism. Was the light once more dispersed? No.

Then – and only then – did Newton admit colour into his account. He noted that the thin beam of light retained not only its shape, but also its hue. A blue ray remained blue, and a red ray remained red, no matter whether he reflected it off coloured surfaces, passed it through coloured media, or added extra prisms. Nothing disrupted the beam's spatial or chromatic integrity.

Whatever property first caused the light to fan out could not, therefore, be a property of prisms. This left only one possibility: he had uncovered a property of light.

Newton tentatively assigned physical values to his different coloured lights. Red light was only slightly deviated from its path by a prism; violet light was turned from its course much more sharply. Oddly, there was no white light in the spectrum. Where had it gone?

When Newton brought the coloured lights to shine on the same spot of wall, white light reappeared. White light was not one thing: it was a mixture of different kinds of light. So what is colour? Newton had, in a sense, found a way to measure colour; he could record how weakly or strongly coloured lights were refracted. But he could not say that blue light refracts by such-and-such amount because it is blue. This left him with the job of somehow separating the physical properties of colour from their phenomenological properties.

He failed. Worse, he knew that he had failed. His alibi – he claimed his pet dog Diamond knocked over a taper and set light to papers containing his best thoughts – is so desperate one is almost inclined to believe it. More telling, perhaps, is the fact that he delayed publication of his *Opticks* by twelve years, by which time his fiercest continental critics were safely dead.

Bring the lights of the spectrum together, to shine on the same spot of wall, and they reconstitute white light. Newton was inclined to take this observation at face value, but not every light source has as rich a spectrum as sunlight, and two years later, Huygens pointed out that as few as two colours (blue and yellow were his example) were sometimes sufficient to reconstruct white light. It was a trick Newton himself never mastered: 'I could never yet by mixing only two primary colours produce a perfect white', he lamented.[6] However, he confirmed that white is produced when just three colours, spread reasonably evenly across the spectrum, are brought to a point.

What was the nature of light, if just two or three colours could produce a sensation of whiteness in the eye? Were there really no more than two or three primary colours, from which all other colours derived?

Newton thought of light as a stream of particles of different sizes, each generating a different spectral colour in the eye. Newton's great rival, Robert Hooke, studying the gaudy multi-coloured shimmer of soap bubbles, advanced a very different theory of light and colours, arguing that light was a mixture of rays which came in a number – perhaps an uncountable number – of different wavelengths.

Homely evidence against corpuscular theories of light had already appeared by 1665, in a posthumous collection of optical observations by the Jesuit physicist, Francesco Maria Grimaldi. Grimaldi's close observation of light revealed that light blurred as it emerged from a narrow aperture, just as coastal waves bend when they enter the narrow aperture of a harbour. (He coined the word – diffraction – for this phenomenon.) He lit an object with a single source of light, passed it through a narrow aperture, and noted that when a second, similarly narrow source of light was brought to bear on the same point, this would sometimes cause the point to grow dimmer. We would call this the 'interference effect'. Grimaldi's observations hardly amounted to a wave theory but they were incredibly damaging to any assertion that light was a kind of matter.

How long Newton would have persevered with his corpuscular theory in the face of mounting evidence is anyone's guess. We know that he toyed with a wave theory of light, and it is tempting to suppose that it was chiefly his animosity towards his arch-rival Robert Hooke that persuaded him otherwise.

When, on 12 November 1801, the Royal Society received conclusive proof that their hero Isaac Newton had been in error – that light was after all a wave – their response was unedifying. The hostile reception given to the bearer of these bad tidings, a twenty-eight year old doctor, Thomas Young, was sufficient to stall what up till then had been a most promising career.

Born in 1773 to Quaker parents, the eldest of ten children, Thomas 'Phenomena' Young (so nicknamed by his Cambridge contemporaries) could read fluently by the age of two, and by sixteen was proficient in Latin and Greek and acquainted with eight other languages including Hebrew, Arabic and Persian. Young had been ruffling the feathers of intellectual London since 1793 when, as a twenty-year-old medical student, he showed members of the Royal Society how the human eye accommodates to objects at different distances by using the ciliary muscles to alter the curvature of the lens. In 1797 an uncle left him £10,000 and a London house, into which he moved in 1800. As a man of independent means, Young devoted his spare time to investigations into sound and light, with contributions so frequent and numerous that he sometimes had to invent pseudonyms to avoid the charge of neglecting his medical work.

By 1801 Young was professor of natural philosophy at the Royal Institution and one of its first lecturers. The RI's public talks became famous for their hazardous stagecraft: early audiences regularly contended with toxic fumes, safety lamps plunged into explosive gases, powerful electromagnets dangled above their heads, and model volcanoes.[7] Young's demonstrations were, by comparison, relatively staid. However, what he lacked in showmanship, he more than made up for in diversity. In 1801, he lectured on acoustics, air pumps, animal life, astigmatism, and astronomy (to cover only the first letter of the alphabet). 1801 was also the year he showed that light was a wave.

'The experiments I am about to relate,' Young assured his audience (at a later and fuller demonstration, this time in 1803), 'may be repeated with great ease, whenever the sun shines, and without any other apparatus than is at hand to every one.'[8] He split a narrow beam of sunlight with a slip of card, about one-thirtieth of an inch thick. The card, held edgewise to the tiny beam of light whose diameter was only slightly greater than the thickness of the card, split it into two slivers, one on each side. These famous demonstrations – a refined version of Grimaldi's work of some 150 years before – proved that light is a wave: the light bent around around the edges of the card and spread out to produce an interference pattern, just as ripples on a pond interfere.

The Royal Society's old guard – keepers of the Newtonian flame – shrugged and shook their heads. One of the younger members went a step further, and sharpened his pencil to a wicked point. Five years Young's junior, Lord Henry Peter Brougham (1778–1868) also had something of the prodigy about him; he was destined to become Lord Chancellor and a powerful opponent of the slave trade. At the time of Young's demonstration, he was a versatile and prolific writer for the *Edinburgh Review* (founded a year earlier in 1802). He contributed eighty articles to the first twenty issues, ranging over science, politics, literature, surgery, mathematics and the fine arts. Brougham had already presented a modest paper concerning light and colour to the Royal Society in 1795 and an (even more modest) paper on prisms in 1798; achievements which, in his own mind, entitled him to launch a savage, satirical and anonymous attack on Young in the pages of the *Review*.

Young had no appetite for a scrap. He abandoned his researches into light and turned his polymathic mind to other, less controversial subjects; he came up with the first good scientific definition of energy and helped translate the Rosetta Stone.

If violet light had a wavelength of 0.0000016 inches (400 nanometres) and red light had a wavelength not quite double that (about 700 nanometres) what would be the nature of the radiation that lay outside those wavelengths? Might there be such a thing as *invisible* light?

The British astronomer Sir William Herschel (1738–1822) was interested in the relative energies (measured as heat) of differently coloured light. He placed a sensitive thermometer just beyond the red part of the spectrum – and discovered infra-red radiation. Learning about William Herschel's discovery, the German chemist Johann Ritter (1776–1810) set about detecting 'invisible light' beyond the violet end of the spectrum. From his own experiments, as well as from the new lore of photography, Ritter was familiar with the way silver chloride turns black when exposed to light. He also knew that blue light caused a speedier reaction than red light. Using a prism, Ritter exposed silver chloride to an area just beyond the violet end of the spectrum and noted an even more intense reaction. Silver chloride reacted somewhat to visible light, but what really got it going was ultraviolet – a light nobody could see.

By the turn of the nineteenth century, it was clear that whatever correlation there might be between the physics of light and the perception of colour was mediated through biology. Although Thomas Young is still remembered, this has little to do with his disregarded lectures of 1802 and 1803. Auguste Fresnel achieved a larger, better set of observations and proofs only a few years later – testament to his gift for experiment and the intellectual advantages of a quiet home life. (A Napoleonic court, rightly suspecting that he harboured Royalist sympathies, had sentenced him to live with his mother.) The power of his deductive reasoning led Young to the theory by which he is best known today. As James Clerk Maxwell was later to say, in a lecture to the Royal Institution: 'So far as I know, Thomas Young was the first who, starting from the well-known fact that there are three primary colours, sought for the answer to this fact, not in the nature of light, but in the constitution of man.'

The spectrum contains only a few of the colours we see around us. There is no brown band in the rainbow, although the natural world throws up any number of different browns. Light of different wavelengths appears differently coloured to us, but it cannot be the light itself that is coloured. Colour must be the means by which the brain tells us what mixture of wavelengths is being reflected by an object.

How many wavelengths must we see to perceive, in their mixture, a virtual infinity of different colours? Young knew the number could not be large, because a room lit by red lamps, for example, is not very much dimmer than a room lit by white. If the eye had to have many different kinds of colour-receptive abilities, for many different kinds of light, there wouldn't be enough red-sensing ability to see clearly in a red-lit room. So 'it becomes necessary,' Young wrote in 1801, 'to suppose the number limited, for instance, to the principal colours, red, yellow and blue.' With no anatomical evidence, Young had inferred the existence of three types of colour receptor in the human eye.

Young plucked the colours red, yellow, and blue out of the air. From the writings of Newton and Huygens, Young had noted that a mix of any three coloured lights will generate white light, provided their colours are reasonably spaced out across the spectrum. (He seems to have found no use for Huygens' observation that just two well-chosen spectral colours can perform the same trick). But Young was also intrigued by two other observations: that green and red light, shone on the same spot of wall, will generate pure yellow light and violet and green light will generate blue.

Young eventually revised the triumvirate of colours perceived 'directly' by the eye. In his article 'Chromatics' for the *Encyclopaedia Britannica*, he wrote that the eye detects mixtures of red, green, and violet light, manufacturing from these colours all the other colours we can see, even such apparently simple and 'primary' colours as blue and yellow.

Flooded with light, cones become chemically exhausted and need time to recharge. Deprived of light, cones become hypersensitive. Just as staring at a black dot for thirty seconds or so can leave a bright dot persisting in your vision for more than a minute, so staring at strong colours can upset the eye's colour balance.

Consider the image of the Union Jack in the colour section. The expected colours have been swapped for their complementaries. A complementary colour is the colour which, when added to the original colour, generates white light. A colour and its complementary will do this because under their mixture all cones are being stimulated equally.

In the figure, the red parts of the Union Jack have been swapped for cyan (the blue that arises when green and blue light overlaps). The dark blue parts (very close to violet in the spectrum) have been swapped for yellow, the colour made by overlapping red and green lights; and the white has been swapped for black. Concentrate on the dot at the centre of the image for thirty seconds, then switch your gaze to the wall. For a fleeting moment, you will see the after-image of a Union Jack, in its true colours. Why? First, the parts of your retina exposed to black have been rested and recharged; once your gaze turns to the wall, their greater readiness to excitation generates an illusion of bright whiteness. By exhausting the cones receptive to shorter wavelengths of light, the cyan bars of the Union Jack have primed those portions of the retina to perceive red. The yellow, by contrast, drains both the 'red' and 'green' cones, so that when the parts of the retina exposed to yellow view the wall, a dark blue, almost violet, after-image is generated.

The eye is not a painter's palette. The eye gathers and interprets a mix of wavelengths. Add a wavelength and you give the eye an extra morsel of information to work with. The more wavelengths there are, the richer the view. This is why, when shopping for clothes, it's best to take them to the front of the shop, where you can see them in daylight. Daylight reveals more colour information than artificial light, and some artificial lights are better than others. (Fluorescent lights are hopelessly drab.)

On the painter's palette, something different happens. The more pigments are added, the *less* colour information the mix contains. Pigments swallow most wavelengths of light; it is the wavelengths they do not absorb which bounce into our eyes. A jar of turmeric

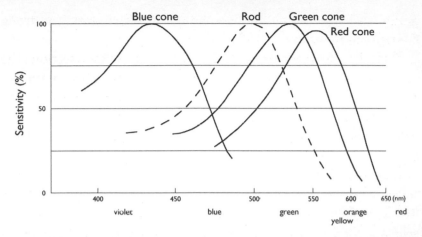

absorbs all light but yellow; a jar of paprika absorbs all light but red. Mix the two and the light reflected will be orange – a mixture of red and yellow light. It won't be a particularly vivid orange, because the yellow turmeric and the red paprika each absorb a little orange light. The more colours we mix in, the duller the result: eventually all wavelengths of light will be absorbed by at least some of the mixture. Strictures against the mixing of pigments are a commonplace of painting manuals.

These two perfectly correct and completely incompatible ways of looking at colour – subtractively like a painter, or additively like a projectionist – can trip up the keenest. David Brewster, researching colour vision in the 1840s, made the mistake of treating coloured lights as paints on a palette. Instead of shining lights through different coloured filters on to a single point, thereby adding monochromatic lights together, Brewster shone light through filters placed on top of each other. Add red light to green light, and you get a vivid yellow. Place a green filter under a red filter, however, and the little light that penetrates both is a darkish red.

Brewster managed to convince himself that each part of the spectrum was made up of three primary colours: yellow, red, and blue. He persisted for years in this eccentric opinion, which seems odd, given the precision of his observations. Thinking he was measuring the relative power of his three 'primary' lights, Brewster inadvertently measured the relative impact different colours have on the eye, and so produced what, twenty-five years later, Helmholtz realised were the first really good drawings of how different cones react to different wavelengths of light.

Brewster's diagram clearly revealed the wavelength sensitivities of our three types of cone: violet on the far right (which Brewster called blue), green in the middle (which Brewster, for some reason, reckoned was yellow) and on the right – orange. This is easily the diagram's biggest surprise and one Brewster never spotted: there is no cone attuned to red light. The best our so-called 'red' cones can capture, according to Brewster's diagram, is a yellowish orange. Later assays, like the one on the previous page, produced readings in which the long-wave pigment did not even peak at orange, but at yellow-green!

If we are surprised to learn that blue and yellow are perceived as mixtures, how much more startling it is to discover that red – most vibrant and unsettling of colours; the colour of danger, blushes, poison, blood, and sexual arousal – is not directly perceived, but constructed whenever yellow cones are excited and blue cones fall silent. Indeed the sensation of red can be generated by removing all bluish wavelengths and enriching the yellowish-green part of the spectrum. The fact that red is patently more than this formula is striking evidence that colour is an neither an objective property of objects nor of the light they reflect: it is a construction of mind.

This is not the only surprise the diagram contains. Brewster had every reason to expect that the curves on his diagram would exhibit a tidy symmetry. He found, however, that there were modest but significant variations in the relative strength of his colours as he moved across the visual spectrum. Accordingly, Brewster nudges his violet curve to the right of the diagram, while his red and yellow curves run more closely together. This first-class observation is but a hint of what would be revealed by later researchers, using more sophisticated equipment. Today's benchmark measurements create a picture even less symmetrical than Brewster's. The 'green' and 'orange' curves run even closer together, so that they are practically overlapping, while the 'violet' curve sits out on a limb, detecting wavelengths of much smaller wavelengths. This far-from-symmetrical arrangement suggests that while blue cones detect wavelengths far into the shortwave band of visible light, red and green cones both respond to very similar long wavelengths. Couple this with the fact that the three sensitivity curves overlap, so that even the simplest-seeming colour is detected by at least two types of cone (light between violet and green will stimulate all three types!) and one begins to wonder whether it makes any sense at all to talk about cones detecting this or that 'colour' of light. It would be a shame to eschew talk of 'red', 'green' and 'blue' cones: there is a certain prettiness and poetry to the habit. But there is no anatomical justification for it: red, green, blue, nol, wor, wheatmeal, and every other colour we see (except for extreme violet, detectable by blue cones only) is detected by at least two types of cone.

So far, we have talked about lights. No one has mentioned shadows. We have talked about the colours of the spectrum. No one has mentioned earth colours – the browns and olive greens that light alone cannot create.

Otto von Guericke is one of those people footnotes were invented for. His genius is hard to pin down. He invented a machine for the production of static electricity but he did not realise what it was. He invented a barometer and a vacuum pump, and he directed the rebuilding of the city of Magdeburg, razed during the Thirty Years War in 1631. Most importantly for us, in 1672 he wrote the first ever methodical account of the colour of shadows.

Light an object with a red light: it appears to cast a green shadow. Light the object with a green light, and its shadow appears red. Shine a yellow light, and a blue shadow is cast. A blue light casts – well, one wants to say a 'yellow' shadow, but oddly, there is no such thing; instead the shadow appears brown. It is one of the foibles of the eye that it perceives dim yellow as a separate colour. (There is also, no doubt, a linguistic element to the way we now define yellow so narrowly: in the eighteenth and nineteenth centuries, 'Indian yellow' paint, which used a pigment made from the dung of cattle fed on poisonous mango leaves, lent European interiors a hue that to modern eyes seems more brown than yellow – not to say rather diarrhoetic.)

Dazzled by brightly lit patches of one particular colour, the brain is somehow being tricked into filling the shadows with a complementary colour, just as, a couple of pages back, it was tricked into constructing a true-colour Union Jack on a white wall. The difference

– the crucial difference – is that we can explain the appearance of the true-colour Union Jack by supposing that our colour receptors are exhausted by prolonged exposure to lights of certain wavelengths. However, shadows cannot be explained in this way. A strong colour in one part of the scene should not, by Young's theory, upset the colour balance of receptors directed at quite another part. And our experience of coloured shadows is quite different from our experience of after-images. After-images are blurry and float about in our vision: coloured shadows are vivid, precise, and fully integrated with the scene before us. This observation reveals a whole new order of colour; a system of antagonistic pairs, in which the absence of one colour brings forth the sensation of another, so that blue 'opposes' yellow, and red 'opposes' green.

Ewald Hering, a German physiologist and psychologist who taught at the University of Leipzig, came up with this formula in 1872. Hering wondered, with devastating insouciance, why certain colours should be unimaginable. While it is perfectly easy to imagine a greenish yellow, or a bluish green, or a yellowish red, it is considerably harder to think of a 'bluish yellow', or 'reddish green'. Hering mapped these 'impossible colours' over the spectrum and deduced hitherto unknown relationships between the colours. According to his scheme, the eye was sensitive to four wavelengths, not three, and these sensitivities were arranged in opponent pairs: blue and yellow, and red and green. Part of Hering's supporting evidence was the observation that people are never colour-blind for a single colour. Invariably, they are missing one or other of his opponent pairs: those who are colourblind to red are likewise blind to green, and the few who fail to see blue also fail to see yellow. White and black arguably made a third pair, bringing the number of Hering's primary colours to six.

There was genius in including white and black among the primaries. By introducing light and shade into his discussions, Hering was able to explain how we see earth tones. The eye does not reveal absolute values for how bright or how dim things are in a particular scene. Absolute light values are almost meaningless. Place a cube of coal on a table and shine a spotlight on it. Beside it, in the deep shadow beyond the circle of light, place an identical cube of chalk. Both coal and chalk shine with the same vigour. Chalk readily reflects what little light reaches it; coal reflects the brightest light only with reluctance. To see these objects for what they are, what matters is that the coal is bathed in light, while the chalk lies in shadow. The eye ignores the fact that the coal and the chalk glow with the same intensity: it points up the local differences in illumination; the white glare surrounding the coal, and the deep darkness around the chalk. No one has ever confused coal with chalk.

Only if we rob an object of all context can the eye be fooled. A full moon at night, lit by a sun we cannot see, and suspended in the non-reflecting vacuum of space appears white – yet moon-dust is black.

Hering realised that the eye renders light and dark by comparing the light levels in neighbouring regions of space. This awareness enabled him to describe how browns and olive greens arise, even though no mixture of lights in a darkened lab can generate them. When yellows and greens are surrounded by areas of greater illumination, browns and olives appear; they are seen as mixtures of black with either yellow or green. To interpret these colours correctly, we need a context. Look at a brown or olive surface through a long, non-reflective black tube and you will see either orange, yellow, or green. The dun, 'earthy' quality of the colour is quite stripped away.

Given that Hering had managed to explain the perception of common, yet hitherto mysterious colours, it seems strange that his work should have encountered resistance. The problem was that Hering's idea of opponent pairs sat uneasily with Young's explanation of colour-blindness. Young maintained that some people were missing one out of three colour-perceiving 'particles' (or, as we would say now, one out of three types of cone cell.) This might not have mattered – three-quarters of a century had passed since Young's paper, and few remembered it – but his theories had recently acquired a new champion in Hermann von Helmholtz, the most influential vision scientist of the nineteenth century.

Helmholtz's most lasting achievement in the study of vision is his *Handbuch der Physiologischen Optik* – a work which required frequent updating in his own lifetime, not least because of the sheer number of arguments and objections Ewald Hering put up against

Hermann von Helmholtz.

it. The two men could agree on nothing. Their arguments ranged across vision research, from spatial perception to eye movements, stereoscopic vision, and, finally – with Helmholtz retired from research and reportedly sick of Hering's bickering – to colour vision.

Helmholtz was immovably wedded to the theory of trichromacy. He came up with the idea independently, only later – by happy accident – discovering Young's neglected paper. Young's arguments, re-aired, confirmed the rightness of Helmholtz's trichromacy idea to everyone's satisfaction – everyone's but Hering's. Hering's theory of opponent pairs flatly contradicted both Young and Helmholtz.

Although Hering is often cast as the iconoclast in accounts of the ensuing argument, it was he who came closest to resolving the dispute. It struck Hering that he and Helmholtz might both be right; that they had described different *stages* in the perception of colour. Perhaps, as Helmholtz said, there were indeed just three kinds of colour-sensitive receptor in the eye. In that case, Hering's 'opponent pairs' were not *structures* but descriptions of the activity of the three receptors. When the receptors were excited, red, yellow, or white were perceived. When the receptors grew less excited, one perceived their opponent colours: green, or blue, or black.

Hering's attempt at a *rapprochement* got nowhere. His theory and the Young-Helmholtz theory were at loggerheads for nearly a century. Each had its champions, and each side was convinced theirs was the 'correct' explanation of colour vision. The debate was heated, endless, and in some cases, downright unpleasant. Even today, vision theorists fall into two broad camps: those who reckon Hering was a visionary and those who reckon he was a lucky eccentric.

What was so extraordinarily difficult about Hering's ideas? First, he proposed that a reduction in nervous activity might lead to a sensation of certain colours – green, blue, and black – just as readily as an increase in activity would lead to the sensation of red, yellow, and white. But how could reduction in nervous activity give rise to sensation? He did not know. Next, he proposed that to perceive the full gamut of colours – including olives and earth tones – the eye integrated information from neighbouring parts of space. How? No one knew.

What Hering needed was a good model of the nervous system and there wasn't one. There wouldn't be until well into the following century. Hering died in 1918, when Haldan Hartline – who unpicked the mechanisms of contrast – was a teenager, and Edwin Land – who realised Hering's ideas in mathematics and technology – was just nine years old.

3 – Old wine in new bottles?

Ed is some star in his studies and we are sure that he will make a name for himself and Alma Mater in college.
Caption to a college year-book photograph of Edwin H Land

After Thomas Edison, Edwin H Land was the most prolific inventor ever to do business with the US Patent and Trademark Office. By 1982, when he retired from Polaroid, the corporation he had founded in 1937, Land had over five hundred patents and a personal fortune, and left a legacy in vision research that remains controversial. Did Land's 'retinex' theory of colour vision really advance our under-standing – or was it simply a repackaging of old ideas, some of them going back as far as the eighteenth century?

Edwin Land was born in Connecticut in 1909, the only son of a prosperous scrap-metal dealer. Optics fascinated him from boyhood; he was only nineteen when he invented the sheet polariser which launched his extraordinary career. His energy and determination were legendary. Making his first polarising filter required the use of a gigantic electromagnet, which was kept under lock and key in Columbia University's physics laboratory. Refused a key, Land went to the sixth floor, climbed on to a ledge, and edged his way along the side of the building to gain entrance to the lab through an open window. After a year of this sort of thing, Land decided to continue his education at the New York Public Library.

His career was not without its checks and disappointments. For decades, he argued that the most important application for his polarising filter would be as an antiglare device for car headlamps. The car manufacturers did not listen, and late in the day they came up with a cheaper solution: dipping head lamps. Land, who had expected to save lives, instead benefited humanity in more indirect ways: he made sunglasses, 3D films, innumerable laboratory and industrial instruments, and, one sunny day in 1944, while wandering through Santa Fé zoo with his daughter, he dreamt up polaroid photography.

As a part of his research into instant colour film, Land repeated some of James Clerk Maxwell's experiments.[9] He was particularly interested in Maxwell's attempt, at the Royal Institution on May 17, 1861, to produce colour images using three-colour projection. Maxwell had taken three black-and-white transparencies of a tartan ribbon: one through a red filter, one through a green filter, and one through a blue filter. He used magic lanterns to superimpose his transparencies on a screen, filtering the red image with a red filter, the green image with a green and the blue with a blue. The experiment was a success, producing 'a coloured image ... which, if the red and green images had been as fully photographed as the blue, would have been a truly-coloured image of the ribbon.' (Maxwell knew that the photographic materials of his day reacted rather unevenly to different wavelengths of light.)

One evening, at the end of a long series of experiments with three projectors, Land and his assistants shut off their blue projector and took the green filter out of the green projector. Then, one of Land's assistants, Meroe Morse, called their attention to the screen. The red projector was still running, projecting the red record on the screen in red light, and the unfiltered green projector was projecting the green record with white light. That combination of red and white lights should, in Morse's mind, produce something pinkish. But there was the original image, its every colour still identifiable. How could red and white lights throw blues and greens on the screen?

Land was not particularly impressed. The complementary colours thrown by shadows had been known to von Guericke four hundred years earlier, and colour opponency – the greenish tinge given to objects adjacent to red-lit surfaces, the bluish cast of shadows cast by yellow lights, and so on – was one of the mysteries of colour perception with which Hering had twitted Helmholtz. Certainly, explaining the mechanism of colour opponency had been a problem for Hering, but Haldan Hartline had demonstrated how nerve cells in the retina detect contrast by mutual inhibition. So it was generally assumed that opponency must involve inhibitory structures like the ones Hartline had discovered.

'Oh yes,' Land explained to Morse, 'that's colour adaptation' – and they went home.

At two o'clock the following morning, Land sat up in bed. Colour adaptation? What colour adaptation?

If colour adaptation were a simple matter of opponency, the most they could have expected of that red-and-white image was a few greenish edges here and there. But the image cast on the wall had been better than that – much better – it contained a full spectrum. Somehow their eyes had deduced, from the little evidence on the screen, the complete spectrum of the original image. How was a mystery, but Land, considering what information the eye had to work with, was able to sketch the broad outlines.

First, the eye had information about the reds in the image. This not only told it what objects were predominantly red; it also revealed what proportion of red was contained in the colours of all the other objects. Second, the eye had information about the greens in the image. The trouble was, there was no way the eye could know this. The green filter had been removed from the projector and the green record was projected in 'noisy', multiple-wavelength white light. The eye was receiving information, not about 'green' light so much as information about light that was 'not very red'. Nonetheless, the contrast between red wavelengths and white illumination was enough for the eye to manufacture an entire spectrum.

Most colour-vision experiments up till then had used single objects or swatches of a single colour. These experiments had been uninformative, and now Land knew why: the experimental set up had reduced the eye to a mere wavelength detector. Place a swatch of isolated colour under a green lamp and it appeared green. Placed under a red lamp, it glowed red. Big deal. It was only when objects were put into context that the eye could deduce colours, the way it had that night in Land's lab.

Land's 'retinex' theory of colour vision arose from a series of dramatic experiments, designed to see how far colour perception could be stretched. Land's team put colours in context: they used boards of intersecting shapes and colours, called 'mondrians' after the artist whose work they resembled. People studying the mondrians reported colours accurately, even under the most extreme changes in lighting. For instance, by adjusting the lighting, a swatch might give off a greenish wavelength one moment and a reddish wavelength the next. If the swatch was seen in isolation, this was what was observed: the swatch changed from green to red. When viewed as part of a mondrian, the object's colour remained stable.

Context is everything. The eye has no interest in absolute levels of illumination, nor in absolute colour values. Every colour is perceived in relation to every other, just as every patch of light is perceived in relation to every patch of shade. In the real world, this makes colours remarkably stable. Reduce the visual context, and we become susceptible to illusions of contrast like the squares in the colour section. The three squares in this figure are particularly interesting, since they demonstrate three different contrast effects at work, affecting our perception of lightness (how bright a colour is), hue (what colour it is), and saturation (how easily it can be distinguished from a grey of the same lightness).

Until then, researchers had rather assumed that colour progressed in a paint-by-numbers fashion, 'filling in' a black-and-white sketch point by point. Now it was clear that colour perception used all three aspects of colour – its lightness, its hue, and its saturation – to detect boundaries and sketch forms. Colour vision was not so peculiar; it was simply a multi-dimensional version of monochrome vision.

In the colour section, Mach's bands are adapted to reveal how we perceive hue and saturation. The 'fluting' effects, though milder than those generated by the original, demonstrate that hue and saturation are, like lightness, processed by a contrast-enhancing mechanism.

Land's own demonstrations were spectacular. Using two yellow lights with wavelengths only 20 nanometres apart – the difference is barely detectable to the eye – he was able to create full-colour images.

Ironically, Maxwell's original 1861 experiment had done much the same but without his knowing it. Though Maxwell appreciated that the photographic processes of his day registered wavelengths unevenly, neither he nor anyone else realised quite how insensitive they were. In truth, Maxwell's 'red record' and 'green record' contained no unique information, as his film stock was only capable of registering blue light! Not until 1873 was there a photographic emulsion sensitive to any light but blue. Good responses to the whole visible spectrum were not achieved until 1882. The history of science is full of these curious footnotes: the truly 'original' experiment is rare indeed, and every discovery seems, with hindsight, to be but the final piece in a puzzle that has been put together over many years by many people.

Edwin Land had so many precursors – Maxwell and Hering among them – that it seems only reasonable to ask, what did he really achieve? His retinex theory stated that colour perception depends strictly on the neural structure of the human visual system: colour is subjective. Isn't this exactly what Isaac Newton said? Retinex, incidently, was coined from the words 'retina' and 'cortex', because Land did not know – any more than anyone else – whether the retina or cortex had the key role in colour perception. But wasn't this the very question explored by Hartline and his successors?

To put it bluntly, what was original about Land's theory? The mathematics he drew up to explain it were tremendously useful, making retinex a true *theory* of colour vision, whereas everything that had gone before was, strictly speaking, only a description. None the less, many academics, nettled by Land's showmanship and annoyed by his lack of reference to earlier work, considered him a mountebank who was pouring old wine into new bottles. This reaction, though understandable, was not really fair. If journalists wanted to cast Land as an iconoclast, overturning 'establishment' views on colour, it was hardly in his power to correct them. Land was an historian of vision theory, and the lack of cited sources in his papers probably reflects a lack of patience for academic niceties more than an attempt to hoodwink his audience.

His practical work, his algorithms, and even his basic kit – the mondrians and projectors – have become the staple of laboratories dedicated to the development of better cameras and even artificial vision. Edwin Land died, aged 82, on 1 March, 1991.

4 – Fusion and attention

Some say they see poetry in my paintings; I see only science.
                                                    Georges Seurat

Draw the following figure on a piece of card, stick a pencil through the centre, and spin it anticlockwise, not too fast.

The effect is modest, but enchanting: before your very eyes, the card takes on all the colours of the rainbow. The outermost lines are red, the next ones are green, the next pale blue, and the innermost lines are dark violet. Flicker colours were discovered in 1826 by a French monk, Bénédict Prevost, but proved to be one of those puzzles

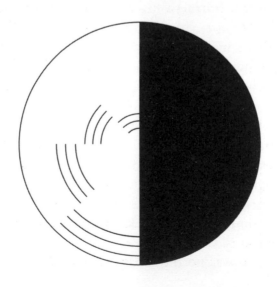

that no-one quite knew what to do with. The phenomenon has been forgotten and rediscovered at least a dozen times, most famously by the toymaker C E Benham, whose odd little top sold well and immortalised his name into the bargain. Nobody knows how Benham's top works, or how significant it is. It could be the key to unlock a radical new idea about colour perception – or it could simply be a case of the overloaded brain getting its wires crossed.

The brain quite often treats seperate streams of information as a single, anomalous input. Here, courtesy of Hermann von Helmholtz, giant of German natural philosophy, is another toy to confound the philosophically minded. Using a circle of card and a pencil, make a spinning top. Mix blue and yellow paint to make green, and paint the mixture over the centre of the disc. Paint half the outer ring blue, half yellow. When you spin the disc, the centre looks – well, green, obviously – but the circumference looks lighter, and grayish.

In an age where coloured lights were awkward to project and disco had yet to be invented, spinning a coloured top like this was a good, cheap way of demonstrating the difference between coloured lights and coloured pigments. Different lights mixed together make white, while a mix of different pigments makes a dirty mess. But Helmholtz's top demonstrates something else: when a point in space changes colour rapidly enough, the colours combine in the eye. Blue and yellow, spun together, make white (or at least grey: the light reflected isn't bright enough to make a true white.) Where has this white come from?

An object flickering yellow and blue will excite different colour receptors, and signals from these receptors will be sent off at different times, linked to the rate of flicker. But there comes a point, as the rate of flicker increases, where the signals arrive so close together that the brain treats them as arriving simultaneously. Photoreceptors respond quite slowly to changes in light intensity: locust photoreceptors, for example, need 25 milliseconds to report a change in light – and insect receptors are typically faster than human ones. A spinning top presents the human eye with a rate of flicker it is simply not equipped to handle, so blue light and yellow light blend to white.

Helmholtz's top was no more original than Benham's. It was a demonstration, not a discovery. The principles of visual fusion were already established, at least among artists, by 1731, when the Amsterdam miniaturist turned entrepreneur, Jacques Christophe Le Blon, embarked upon yet another of his ruinous manufacturing schemes. Le Blon, constantly on the look-out for ways to churn out cheap pictures for the burgeoning middle-class art market, had already lost his shirt on a system of three-colour printing. Undeterred, he turned to the textile industry. Weavers were certainly up to the task of producing complex designs: what made woven pictures so prohibitively expensive was the number of different coloured threads used in their manufacture.

So why not use three threads, each dyed a primary colour? Woven finely together, might they not conjure up a full spectrum? Le Blon's experiments were encouraging. Seen from a distance, the colours of adjacent threads fused, as the colours on a spinning top merge to produce novel colours. In both temporal and spatial fusion, the visual system is unable to distinguish between several inputs and treats them as one.

Le Blon was canny enough to realise that this sort of colour mixing was quite unpainterly. Rather than mixing colours, reducing the amount of colour information in the mix, he was laying differently coloured threads side by side. Like differently coloured beams of light directed at a single point, the different-coloured threads, fused in the eye, provided ever-richer colour information. In theory, red, yellow and blue threads, woven tightly enough together, would be perceived as white. The problem was illumination. The primary-colour threads could never conjure up a perfect white, because they were simply not reflective enough; the best he could achieve with them was a 'light cinnamon'.

Le Blon's list of threads increased to four, because he needed black thread to generate different shades of a single colour. With the necessary but philosophically untidy addition of white, the list came up to five. And it kept growing. Dr Cromwell Mortimer, reporting to the Royal Society explained, 'though he found he was able to imitate any picture with these five colours, yet for cheapness and expedition, and to add a brightness where it was required, he found it more convenient to make use of several intermediate degrees of colours.'[10] So much for slashing the cost of manufacture: the illumination problem put paid to any hope of mass-producing pictures in textile form. Le Blon had won himself a permanent place in the history of vision; but the business failed.

Le Blon's story ends happily: Louis XV, impressed with his tenacity, gave him a monopoly to develop colour printing in France. Finally, Le Blon came up trumps: the four-colour method of printing we use today was his invention. He died in 1741, aged seventy-four, four busy and successful years into the project.

Le Blon, a professional painter, knew the value of harnessing illumination to generate colour. Apart from the commercial possibilities of mass production, a painting system that juxtaposed a handful of colours to generate a full spectrum would, by its very nature, generate a uniquely luminous art.

In the mid-1880s, a young painter and print-maker, Paul Signac, ran into Georges Seurat at the newly established *Salon des Indépendants*, an open exhibition which Signac had helped set up in reaction to the art establishment of the day. Seurat was exhibiting *Bathers at Asnières*, which had been rejected by the official Impressionist salon. *Bathers* was composed entirely of dabs of colour. The colour contrast between neighbouring dabs was often extreme; however, the dabs were so small that, seen from a distance, the colours appeared smoothly blended. Another arresting quality was the canvas's comparative evenness of illumination. This was a painting in which shadows seemed to glow as brightly as the lighted surfaces. No wonder the official salon found the work defective; even Renoir – no stranger to the pastel end of the paint chart – found that Seurat's output set his teeth on edge: 'Imagine Veronese's *Marriage at Cana* done in *petit point!*' he complained.

In composing his paintings along the same lines as Le Blon's tapestries, Seurat ran into many of the same problems: the technique does not allow for the fact that, just sometimes, black is black and white is white. For Seurat – an experimental artist in charge of his choice of subject matter – these limits were part of his signature. He might not be able to emulate Veronese with this technique, but then Veronese, master of *chiaruscuro*, could never have emulated *him*.

Signac was impressed enough to begin to explore the possibilities of making paintings out of swatches. He also wrote about the technique, and gave it a name: Divisionism. Though it trips off the tongue less readily than *pointillisme* (which, strictly speaking, applies only to Seurat's style of painting), Divisionism is the proper term for any picture that relies for its effect upon the way colour information from many small areas will fuse in the eye. When you next sit down in front of the television, with its matrix of red, blue, and green dots, you will be sitting in front of a Divisionist picture. Of course, the dots on a television screen are so small that you would have to press your face up to the screen before the picture broke up into individual patches of colour. The canvases of Seurat and Signac were not so subtle – nor were they meant to be.

Bathers is presently in the National Gallery in London. At a distance of about a metre you can focus your attention on any part of the canvas and examine the individual dots. If you want to appreciate how these dots fuse, refocus – not your eyes but your attention – and the colours will blend to reveal the painting as a whole. You do not need to 'zoom'; you do not need to track in towards your subject and out again as though you were the camera in a particularly tricksy Alfred Hitchcock dolly shot. All you have to do is *will* the change from one zone of attention to another.

There are more than forty different types of nerve cell in the retina, each with their own 'field of view'. There are cells which edit the responses of the photoreceptors and bipolar cells that compress that information by factors of between twelve and one thousand, depending on the light level; no one knows how this adjustment is made. The bipolar cells pass their information on to the six layers of ganglion cells, each layer connected with the others by at least thirty different types of amacrine cell. There, the information is once again savagely edited, until only a million signals remain to be carried along the optic nerve to the brain.

Some researchers believe that somewhere in that bewildering mix lurks a mechanism of attention. Perhaps the receptive fields of the ganglion cells – which, although they vary wildly across the retina, always come in overlapping large and small sizes – are part of that mechanism: as we study Seurat's *Bathers*, at one moment we attend to our large-field ganglion cells, and see a pair of purplish swimming trunks; at another, using small-field ganglion cells, we study the painter's individual dabs of blue and red paint. How easy that would be! However, other experiments suggest that we have not just two but up to half a dozen 'zones of attention'.[11]

Colour, which seemed such an intractable phenomenon in the 1950s, now offers us – through the work of artists, printers, and inventors, as much as through the work of academics – a window on to the next great unsolved problem of vision: attention.

No one knows how we focus our attention. No one even knows how to ask the right questions about attention. If the eye is an outpost of the brain, it is – by the same logic – an incursion of the light. When our eyes are drawn to a change in the scene, what draws them: our desire to see, or the change in the scene? Is one a cause, the other an effect? Or – in Goethe's memorable phrase – do 'the two together constitute the indissoluble phenomenon'?

We can no longer talk about the visual organs as any kind of mechanism. That kind of thinking will lead us nowhere, as the German mathematician and philosopher Gottfried Leibnitz pointed out in his *Monadology* of 1714:

Suppose that there be a machine, the structure of which produces thinking, feeling, and perceiving; imagine this machine enlarged but preserving the same proportions, so you could enter it as if it were a mill. This being supposed, you might visit inside; but what would you observe there? Nothing but parts which push and move each other, and never anything that could explain perception.

In the absence of theories – and lacking even the language in which to cast our theories – we have only descriptions. Of all our descriptions of visual attention, Seurat's paintings are, rightly, the most famous. If anyone is so crass as to ask what Seurat's paintings are 'about', the following reply is more revealing than most: they are about the nature of visual attention.

NINE

Unseen colours

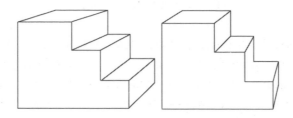

From Manchester, in February 1794, a Quaker schoolmaster, John Dalton, wrote to his friend and mentor Elihu Robinson,

> I am at present engaged in a very curious investigation: I discovered last summer with certainty, that colours appear different to me to what they do to others: The flowers of most of the Cranesbills appear to me in the day, almost exactly sky blue, whilst others call them deep pink; but happening once to look at one in the night by candlelight I found it of a colour as different as possible from daylight; it seemed then very near yellow, but with a tincture of red; whilst no body else said it differed from the daylight appearance, my brother excepted, who seems to see as I do . . .[1]

So, a modest pink flower served as the subject of Dalton's first scientific paper, and launched him on a quite extraordinary career. Dalton (1766–1844), son of a weaver, went on to develop modern atomic theory, formulated several laws on the behaviour of gases, and attained an unlooked-for celebrity for his services to science, philanthropy, and education. Across Europe, even those unfamiliar with his reputation are familiar with his name: *Daltonism* is a common term, in many languages, for colour-blindness.

In the nineteenth century, physiological descriptions of colour-blindness provided clues to human heredity; contemporary genetics has since repaid the favour, by providing us with a rich and complex history of human colour perception. Their fertile conversation is the subject of this chapter.

1 – Dalton's eyeball

Extraordinary Facts relating to the Vision of Colours – Dalton's first scientific paper, published in 1798 – caused something of a stir. Not only was industry making the world more colourful; it was making ever-greater use of colour for its own safe and efficient running. Signals and signs were proliferating, smoothing the relationship between workers and their machines. What could be clearer, after all, than a system of colour signals; a red light for Stop and a green light for Go? Dalton's paper – the first reasonably rigorous description of the nature and prevalence of colour blindness – suggested that colour was not the universal language it appeared to be.

A century on, in the early hours of 15 November 1875, two express trains collided head on at Lagerlunda, Sweden. The authorities were mystified. How could the driver of the late-running northbound express not have seen the red light waved by the station-master at Bankeberg station? The puzzle inspired the physiologist Alarik Frithjof Holmgren to test the vision of the 266 employees of the Uppsala-Gävle railway. Thirteen, including a station-master and an engineer, proved to have difficulties telling red from green. By the end of the year, colour tests had been prescribed for railway and shipping personnel in Sweden, and four year later, in 1879, the American government followed suit, commissioning Dr. William Thomson to devise a colour-blindness test for railway and shipping employees.

Given that the Lagerlunda crash had nothing to do with colour-blindness, this seems a lot of work and fuss. (A revaluation of the original records showed that it had actually been caused by a failure to follow the rules of the railway.[2]) It is not at all clear what tests for colour-blindness have achieved. Those of us who are colour blind are more than capable of compensating for our disability; colour-blind drivers have no trouble remembering that at traffic lights, the red light is at the top, the green light is at the bottom. Anyway, since almost every colour-blind person has at least some sense of colour, practice and experience easily overcome common errors.

The futility of colour testing was shown up most sharply by Mr Trattles, a British seaman denied his First Mate's certificate after failing a colour vision test. Trattles took his case to Winston Churchill, then President of the Board of Trade. His plight was discussed in both Houses of Parliament, and a certificate was finally issued to him after he took a steamer trip, in front of witnesses, down the Thames, and correctly identified every navigation light he passed. But bureaucracy turns slowly. Although the Australian government will let you fly a commercial jet if you're colour blind (and they are the first country to do so), many major airlines still won't employ you. In Britain, restrictions in some branches of the services, the police and civil service are often strongly enforced. And don't even think about driving a train unless you live in Russia. A training course there, using pigment tables, has greatly improved colour recognition among colour-blind railway employees (rather, I suppose, as studying swatch tables has lodged the previously 'invisible' colour indigo in my memory).

Holmgren's curiosity and industry were laudable; but the rapid government action he inspired reflects less a mature measure of risk and more the anxieties of an age accustoming itself to a new form of human calamity: the industrial accident. Today, colour blindness is no more life-threatening than it ever was. It is, though, a significant handicap to anyone whose job involves toiling through the colour-coded arcades of the World Wide Web; too few web-designers remember to make their natty designs friendly to colour-deficient eyes.[3]

To the end, Dalton maintained the modest habits of his Quaker upbringing. He never married, and few men could claim him as their friend. He died, a recluse, on July 27 1844. Nevertheless, when his brethren turned up for the funeral, they found themselves contending with crowds of fellow mourners. A staggering forty thousand people lined up outside Manchester Town Hall for the chance to file past the great man's coffin. The funeral procession was a mile long. Shops and business closed as a mark of respect. Dalton, an active and distinguished public figure, was internationally recognised as the founder of modern chemistry.

The following day Dalton's doctor, Joseph Ransome, began one of medical history's more peculiar post-mortems. Dalton's eyes had not gone the way of the rest of him but lay in a dish on Ransome's table.

Ransome poured the humours of one eye into shallow glass dishes. He noted that neither the aquaeous humour in front of the lens, nor the vitreous humour behind it, contained any trace of colour. The lens was yellow, as you would expect from so elderly an eye. To double-check his result, Ransome carefully removed layers from the back of the second eye and looked through it. Examined through Dalton's dissected eye, red and green objects retained their natural hues.

Dalton, who had single-handedly developed modern atomic theory and transformed our understanding of matter had, thanks to Ransome, reached beyond the grave to conduct his final, posthumous, experiment, a test of the theory he had nursed for fifty years to account for his colour-blindness.

The result was conclusive: he was wrong.

The peculiarities of his impoverished colour vision niggled at Dalton throughout his life. His letter to Elihu Robinson sets out the problem vigorously, describing how the cranesbill in his garden changed colour under different lights. This was peculiar, because others did not see the change. To them, the *Pelargonium zonale* was the same colour, whether swaying in the morning garden or studied by candle-light. Significantly, that colour was pink – a colour Dalton could not see. (He once shocked his co-religionists by turning up to prayer in red stockings – he thought they were grey.)

What caused Dalton to see two colours in a flower when everybody else saw only one? And how was this related to his inability to distinguish green from red?

Thomas Young's theory of trichromacy provided a possible solution to Dalton's problem. In the normal trichromat eye, all three types of cone need to be stimulated equally to see white. Two-colour vision is much more easily 'bleached out' because light has only to stimulate two cones equally to be perceived as white. One by-product of seeing white more often is that objects appear to glisten. (The degree to which Homer emphasises the lustre of objects over their hue lends circumstantial weight to Gladstone's suggestion that the ancient Greeks were colour blind.)

If Dalton was red-blind, Young argued, then a pink flower would have no apparent colour: it would simply reflect ambient light. In daylight it would reflect the blueness of daylight; indoors, it would reflect the light from a yellow candle.[4] Young voiced his opinion – and Dalton rejected it. What set him so firmly against Young's idea? His biography holds a clue.

Dalton was a self-educated man barred, because of his religion, from studying at Oxford or Cambridge. He acquired much of his early knowledge from magazines, and his skills as a teacher were honed by running a school in Kendal, in the Lake District, while he was still only twelve years old. (He constantly had to evade threats of violence from recalcitrant pupils older and bigger than he was.) As a consequence, Dalton never really learned the trick of listening to other people's opinions. This proved, in many cases, to be a key to success: 'Having been in my progress so often misled by taking for granted the results of others,' he wrote, 'I have determined to write as little as possible but what I can attest by my own experience.' Dalton's intellectual self-reliance leant his work an honesty and a novelty that only rarely let him down.

Alas, it also prevented him from ever solving the colour-blindness problem. This talented and original chemist, biologist, and astronomer came up with no better theory than that the humour filling his eyeballs must be blue. The day after his death, Dr Ransome found no evidence to support him, and today, Young's explanation of Dalton's colour-blindness is the one we are most comfortable with.

The patterns of heredity Dalton sketched out in his first paper remained mysterious during his lifetime, but provided valuable clues to succeeding generations, so that colour blindness became the first human genetic trait to be linked to a specific chromosome.

The ophthalmologist Johann Friedrich Horner was born in Zurich on 27 March, 1831, and died there fifty-five years later. A leading campaigner for hygiene and measures against cholera, Horner combined teaching with a crammed research schedule and a busy private practice; over the course of his career he treated a staggering 100,000 patients.

His medical studies introduced him to the ophthalmoloscope, at the time only a few years old, and inspired him to specialise in ailments of the eye. In 1876, he applied his medical knowledge to the mysteries of 'Daltonism'. In place of Dalton's anecdotal findings, Horner presented the first scientific account of the hereditary transmission of Daltonism: colour-blind fathers have colour-normal daughters who are, in turn, mothers of colour-blind sons.

Horner noticed that this pattern had also been described in connection with haemophilia. These ailments were being passed on from generation to generation in the same way, via the mechanism which dictated the sex of a human child – and in fact were firm evidence of the existence of a sex-determining mechanism.

Edmund Beecher Wilson, America's first cell biologist, working at the turn of the twentieth century, was in no doubt that the whole of an animal's development 'is capable of a mechanical or physico-chemical explanation', and Horner's work provided firm evidence in support of this conviction.[5] Wilson asked 'whether the embryo exists preformed or predelineated in the egg from the beginning or whether it is formed anew, step by step in each generation.' Certain developmental decisions – like the sex of the animal – seemed fixed and were relatively impervious to changes in the environment. Those decisions must somehow be stored in the single fertilised egg cell, from which the animal grew.

It had been suspected, for some time, that the 'tiny threads' in cell nuclei, spotted between 1879 and 1882 by the German anatomist Walther Flemming, might be storing instructions for cell reproduction and development. Armed with the latest refinements in optical microscopy, staining, and tissue sectioning, Wilson set about isolating the evidence, and eventually published a paper showing that males possessed one sex chromosome whereas females possessed two. This, Wilson concluded, must be the difference that decides the sex of the organism. Even as the paper went to press, Nettie Stevens – the fore-most female biologist of her day – submitted a paper which flatly contradicted him. Stevens had discovered, quite independently, that males have two different chromosomes, 'X' and 'Y', while women have two 'X' chromosomes.

Stevens was right but so, oddly enough, was Wilson, for he had stumbled upon a rare case. Some males *are* born with only one sex chromosome. Much more frequently, though, they heed Nettie Stevens, and boast two. Stevens had saved Wilson no end of tail-chasing. By 1911 Wilson and his colleague, Thomas Morgan (Stevens's old professor: a fitting coincidence), were able to show that colour-blindness must somehow piggy-back on the X-chromosome.[6]

This is why girls only rarely suffer from colour blindness. The only way a girl can be colour-blind is if both her parents carry colour-blind X-chromosomes. If (as is much more likely) a girl inherits a colour-blind X-chromosome from just one parent, she will still be able to see the full complement of colours, because, as an embryo, only some of her cells will follow the instructions of her colour-blind X-chromosome; others will listen to her second, colour-normal X-chromosome. Although the initial 'choice' of which X-chromosome to listen to is random, tissues instructed by the healthier X-chromo-some tend to dominate the development process (the phenomenon of 'X-chromosome inactivation'). So, a woman carrying one 'colour-blind' X-chromosome will still see the normal complement of colours. Her sons will not be so lucky. A son inherits only one X chromosome, and it must come from his mother. (An X-chromosome from the father would give the child two X-chromosomes, making it a girl rather than a boy.) Since one of the mother's X-chromosomes carries a gene for colour-blindness, there is a fifty per cent chance that he will be born colour-blind.

This is where most accounts of colour-blindness end. But there is more to tell, and in the telling, we get a much richer idea of how human colour vision evolved.

Air-dried by Ransome and donated to the Manchester Literary and Philosophical Society, Dalton's eyes survived the German bombing which destroyed so many of his valuable papers. In 1990, a team of vision researchers obtained permission to re-examine the shrivelled remains.[7] This was, as one of the team, John Mollon, described it, an exercise in 'molecular biography': a sincere (if peculiar) act of homage to establish, once and for all, which of the several common forms of colour blindness afflicted Dalton.

Dalton, it turned out, was not 'red blind' at all. He was green blind. Mollon's team of 'molecular biographers' showed, by genetic analysis, that Dalton's green cones were missing. Where you would expect there to be green cones, there were simply more red cones. This condition is 'deuteranopia', and it is fairly uncommon. It is much more usual for the green cones to be recognisably 'green' but dysfunctional in some way. This gives rise to the milder condition of 'deuteranomoly', an umbrella heading, which covers a number of possible problems, including pigments that react poorly to light, and pigments that react to unexpected wavelengths. There are equivalent conditions that affect the pigments for long-wavelength and short-wavelength light: protanopia, protanomoly, tritanopia, tritanomoly . . . With so many varieties of 'colour-blind' vision, it seems extraordinary that anyone ever acquires perfect colour-normal vision. In fact, few do.

Jeremy Nathans, professor at Johns Hopkins University School of Medicine, has a Golden Brain. This is an award from the Minerva Foundation, a Berkeley-based charity with a special interest in brain and vision research. Received in 1989, the award recognised Nathans's work in unpicking the genetics of colour vision – work which threw a bright light on the many puzzles of colour blindness. Like Walter Gehring, Nathans started his career studying the genetics of *Drosophila*. It was not long, however, before he fell in love with rhodopsin – that ancient, ubiquitous pigment whose slow mutations, read in the right way, are an historical record of vertebrate evolution. Having isolated the gene that codes for rhodopsin in human rods, Nathans went on to isolate the genes that code for human colour receptors. He found that the gene for 'blue' receptors lies in an extremely stable position on chromosome 7, a gene shared equally between men and women. This is why an inability to discriminate blues and yellows is extremely rare, and does not follow the inheritance patterns of the more usual types of colour blindness. Nathans was not surprised to find the genes for red and green receptors on the X chromosome, as it was long established that red-green colour blindness was sex-linked.

With further study, he was able to show why colour-blindness is so common. The genes for red and green receptors are virtually identical (their DNA sequences differ by only two per cent) and lie next to each other on the X-chromosome. When eggs are formed in a female embryo, the X-chromosomes from the maternal grandmother and grandfather exchange sections randomly, to form the egg's brand new X-chromosome. During this process, the 'red pigment' and 'green pigment' genes can easily fall out of sequence. If you imagine the confusion that could reign if two people with the same name move in next door to each other, you begin to appreciate why accidents are common during chromosome assembly. Misaligned and redundant 'red' and 'green' genes account for ninety-five per cent of all variations in human colour vision.

Nathans, who enjoys normal colour vision, was startled to discover evidence of confusion on his own X-chromosome. He had expected to find genes for two receptors: red and green. It turned out he had three. Slips and duplications of this sort are common. Far from there being an easy one-to-one relationship between gene and pigment, it is quite usual for up to nine genes on the X-chromosome to cluster together in an attempt to code for 'red' and 'green'. This is why perception of middle- and long-wavelength light seems to vary even among individuals who see colour normally.

Nathans's work threw up an intriguing possibility. 'Red' genes seem particularly liable to mutation; when more than one 'red' gene is present on the X-chromosome, it often turns out that all the 'red' genes are defective, coding for receptors which register light towards the greenish part of the spectrum. The result, for men, is some degree of colour-blindness. What about women? A woman might have one X-chromosome containing a healthy red gene, while the other contains mutant 'greenish-red' genes. As we've seen, which X-chromosome to 'listen to' is a decision embryonic cells take at random. If there is a developmental disadvantage in following the instructions of one X-chromosome, then cells which obey the healthy chromosome will tend to dominate.

But where is the developmental disadvantage in having an extra colour receptor? Might some women not enjoy four-colour vision?

The first known human tetrachromat is 'Mrs M', an English social worker. She was tracked down by Doctor Gabriele Jordan, then at Cambridge University and now at the University of Newcastle. Jordan found her by testing the colour vision of women who had sons with a specific type of color blindness. Equipped with four receptors instead of the usual three, Mrs M sees rare subtleties of colour. Looking at a rainbow, she can segment it into ten different colours.[8] The rest of us can see only seven – if that many.

Before Mrs M, we thought that the human brain was 'wired' for three-colour vision. We were wrong. The brain can handle however many channels of information the eye cares to throw at it. If our colour perception is so plastic, the idea that the human colour sense has changed markedly during recorded history no longer seems so fanciful.

2 – A history of accidents

We are the products of editing, rather than of authorship.

George Wald

Gordon Walls – a renowned anatomist and expert on animal eyes – was by most accounts not an easy man to get along with. He was arrogant and dictatorial towards his students; of his colleagues, he was demanding. He'd gone in for zoology because, he said, the mathematics in his engineering course became too difficult. Much of what he said about vision was not original, and some of it was wrong. Walls died of a heart attack in August 1962, aged fifty-seven. He is greatly missed.

He was one of vision research's great communicators, as able to write for children as to referee and edit original research. He was also, as one obituary put it, 'kind and warm, but . . . at great pains to conceal it'. For every student he brow-beat in an argument there was one – often the same one – whose casual question he answered with an hour-long lecture of spell-binding clarity. If he was a martinet, he was also a showman, demonstrating night vision experiments by

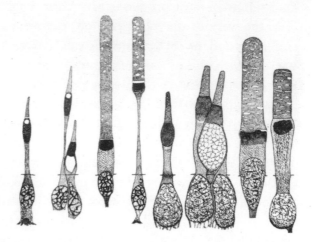

Gordon Lynn Walls' comparison of the visual cells of a leopard frog and a tiger salamander. The four leopard frog cells are on the left.

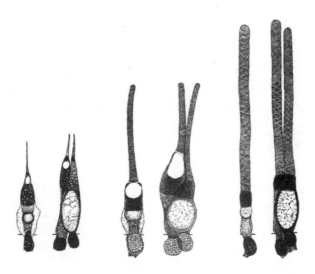

Visual cell types in lizards.

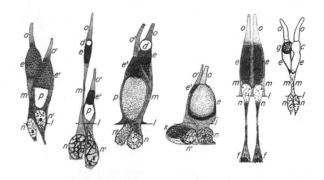

Double and twin cones from fish and reptiles.

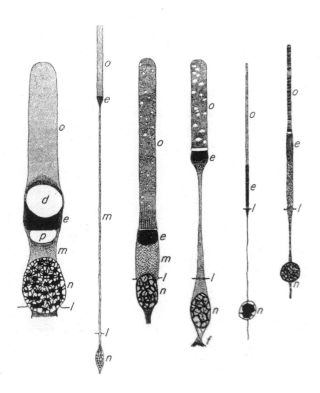

A selection of rod cells; the rightmost rod is human.

moonlight. If he was opinionated, he had a lot to be opinionated about: he could hold an audience for hours on any subject for which he felt enthusiasm. He was once spotted, in a rare leisure hour, conducting impromptu, unofficial tours of the Palomar Observatory.[9]

Walls's stories about eyes drew together the evolution of materials and structures, the rules of optics, and animal behaviour in new and exhilarating ways. He was as fascinated by how eyes moved as by how they handled light; in how they looked, as much as how they saw. He was especially intrigued by the sheer variety of rods and cones to be found in vertebrate eyes. He traced the family traits of different kinds of photoreceptor and was the first to realise that rods are a specialised cell that must have evolved from a primordial cone.

Visual systems adapt to different niches over time; their story is traceable through their genes and – more graphically and entertainingly – through their anatomy. There is the type of frog which, having returned from a nocturnal environment to a daylight one, boasts two kinds of rod – and so has acquired a simple, *ad hoc*, form of colour vision. Then there are the geckos. Geckos are descended from lizards, animals which have spent so long in the sun that they have lost their rods. When geckos adopted nocturnal habits, they had only cones, so geckos are unique in being able to see at night using cones.[10] If only one line of Walls's writings survives into posterity, it will be the title to an essay, a line which handsomely sums up the credo of vision science since Descartes: 'Everything in the vertebrate eye means something'.

Walls's chief written contribution to vision research is the 785-page epic *The Vertebrate Eye and its Adaptive Radiation*, to which he also contributed many illustrations. There is room here for only a tiny sample of Walls's meticulous drawings of the photoreceptors of different species. The further one delves into Walls's catalogue of vertebrate wonders, the more striking nature's variety becomes. There are cones that are tipped with oil. The oil droplet is coloured, restricting the wavelength of light the cone can detect. By tweaking detectable wavelengths this way, many vertebrates enjoy superbly accurate, less 'noisy' colour vision. Many vertebrates – but not mammals.

Many vertebrates have double or twin cones; variations in response between the cones provides instant information about the wavelength

of light. Mammals, on the other hand, have only single photoreceptors.

Many vertebrate eyes adapt to different lighting conditions by
sliding their rods in and out of the pigment epithelium. Mammals
have no such talent.

Most vertebrates enjoy several dimensions of colour vision, and
tetrachromats are common. But not among mammals. Among
mammals, even three-colour vision is a rarity. Most are – from a
human point of view – colour-blind.

Since so many different kinds of vertebrates boast the same visual
equipment – oil droplets, retractable rods and so on – we know this
equipment must have evolved by the time mammals arose, around
240 million years ago. So why did mammals 'throw away' so much
good visual gear?

Every ecological niche is filled by a living thing of a characteristic
build, size, and shape. By studying the fossils of extinct animals, we
can often infer what kind of niche they filled. The earliest mammals
arose around the same time as the dinosaurs. These small, rodent-
like creatures evolved into two groups: marsupials, who give birth
early and nourish the developing embryo in a pouch; and placentals,
who give birth to more developed, capable young. Kangeroos are
marsupials. Humans are placentals. Marsupials evolved to fill daylit
niches, and they have enjoyed stable three-colour vision for over 240
million years,[11] whereas placentals evolved to occupy nocturnal
niches. Humans are the descendants of night-dwellers, and our excellent night vision (and it is excellent, however much people in indus-
trialised countries never get to appreciate it) is one consequence of
our spell in the dark. Another consequence is our relatively gimcrack
and faulty colour vision.

The cobbled-together quality of human colour vision particularly
shows in the anatomy of the retina. You may recall that in Chapter
Four I mentioned the problem of chromatic aberration: the fact that
lights of different colours come into focus at different depths inside
the vertebrate eye. In humans, green and yellow lights, come into focus
at the very centre of the fovea, but blue light reaches the fovea and

surrounding retina in a blurred wash. The human eye handles chromatic aberration by arranging its cones in concentric circles. Blue cones are not found in the fovea but are scattered among the rods filling the rest of the retina.

There is more to this story: first, blue cones are rare, representing only two per cent of all cones. Another thing: look again at the figure on page 227; as orthodox a diagram as you could wish for, which shows the sensitivity of human photoreceptors to light of different wavelengths. The red cone is most excited by greenish-yellowish light; the green cone is most excited by green light and the blue cone is most excited by blue light. This diagram has been copied so often, you may be forgiven for thinking that it is not controversial. But there is a problem.

The vertical axis of the graph measures the percentage of cones stimulated by a particular wavelength of light, or if you prefer, the percentage chance that a particular cone will react to a particular wavelength. It does *not* indicate how powerfully the cones respond. The aggregate responses of blue cones, which represent only two per cent of the total cone population, are nothing like as powerful as the responses of red or green cones. The diagram does not reflect this disparity in signal strength, something which is often overlooked.

Blue-cone signals are maddeningly few. George Wald never managed to detect any, and this was a man whose Nobel prize recognised his work in identifying cones by spectroscopy. Their weakness has driven at least one contemporary writer on vision – Gerald Huth, formerly of the University of Southern California – to dismiss their existence entirely, and construct, in their place, an entirely original (and often persuasive) theory of colour perception.[12] A more conservative interpretation is that our brains must be paying especial attention to 'blue' signals.

The idea that blue-cone information was being handled as a 'special case' was given a boost in 1998, when a paper in *Nature* reported that the blue-cone signals in Macaque monkeys are sluggish – running about 30 milliseconds behind signals from red and green cones.[13] Further studies uncovered evidence that there are indeed two separate channels for colour information – a red-green and a blue channel – which are anatomically separate. In the foetus, they develop in different ways and at different rates; their tissues don't even share the same immune responses.[14] The obvious implication – that they developed independently – is irresistible. But can we prove it?

In 1986, Jeremy Nathans found firm genetic evidence of human colour-vision's split nature.[15] The gene for blue cones is ancient: it predates the entire mammalian line, and is common to many different vertebrates. Establishing the age of red and green cones was more problematic. A survey in 1981 found that all vertebrates have a 'reddish-greenish' receptor of some sort, and the gene that codes for this may be, in many cases, as ancient as the gene that codes for the blue cone. But this gene has mutated many times; mammals, in their reacquisition of colour vision, may have acquired, lost, and reacquired dimensions of colour vision throughout their evolutionary history.

The history of our own species is a case in point: one of our ancestors managed to duplicate its 'greenish-reddish' cone so as to acquire separate classes of 'red' and 'green' cone. We can date this event – the birth of primate trichromacy – by comparing the anatomies of living species on different continents.[16] The primates of South America, which broke away from Africa around sixty-three million years ago, have only two genes to code for cones. Their blue cone is analogous to our own; the other is a 'greenish-reddish' cone. So, the appearance of green and red cone types must have taken place more recently than 63 Ma. More abstruse genetic studies suggest that primate three-colour vision is younger still; just thirty to forty million years old. In evolutionary terms, that is a last-minute job. No wonder it goes wrong so often.

Human colour perception is not the single, seamless sense it appears to be, but the detritus left behind by an untold number of genetic accidents. Our plastic and sense-hungry nervous system wires these mismatched parts together into a single, often imperfect, machine.

The 607 islands of Micronesia are spread across one and a half thousand square miles of the Western Central Pacific Ocean. There are no mountains, or none worth the name; their highest point is the volcano Agrihan which, at under a thousand metres, is somewhat lower than Snowdon, in Wales. Most of these very beautiful islands are little more than scraps of coral, barely breaking the waves that surround them. In a big enough storm, some of them fail to do even that. In 1775, a thousand people were living on the island of Pingelap when Typhoon Lengkieki engulfed the island. Nine out of ten islanders were killed outright; almost everyone else starved to death. Only twenty survived. Little by little, the island and its people recovered, and although intermarriage between close relatives was, of necessity, common, four generations lived and died on the island without suffering any obvious medical problems. So, it was a much-recovered community that suffered the final and oddest consequence of the typhoon. Children began to be born without a sense of colour. Their world was monochrome; a fuzzy, easily dazzled matrix of greys. With time, the trickle of cases became a flood, with each succeding generation producing more and more totally colour-blind (achromatopsic) children. Today, between five and ten per cent of the island's population of 3,000 are unable to see any colours at all.

By tracing the islanders' genealogies, we can see what went wrong. One of the original survivors of the typhoon, the island's young chief, replenished the island with children, but he passed on a rare genetic defect. In the normal run of things such defects are almost never expressed normally, and certainly the young chief would not have been aware of any problem. However, in Pingelap, intermarriage between his descendants spread the defect throughout the population so that, with each new generation, the chances of developing the defect astronomically increased. Yet the Pingelap islanders, afflicted with total colour blindness and photophobia, are far from devastated by their condition and continue to flourish. They have repopulated their island, digging their disability ever deeper into the gene-pool with each passing generation. Today, one in three islanders are, like their illustrious forefather, carriers of the condition.

For humans, the evolutionary pressure to maintain colour vision is decidedly slack. Colour blindness affects many of us – but we do not seem greatly inconvenienced. Eight per cent of males are born colour-blind, which suggests that colour vision is something that our species can learn to live without. The history of human colour vision, as it emerges from researches into our genome, seems to have more to do with the vagaries of genetic mutation and mistake, than with any grim narrative of do-or-die adaptation.

3 – Inside Leibnitz's mill

There are two, virtually independent, pathways for colour information, and they both terminate at the brain's first major hub for visual information, the lateral geniculate nucleus (LGN). The LGN treats the information it receives much as the horseshoe crab's retina treats light. For the LGN, contrast is everything. It compares the signals from red cones to the signals from green cones, and enhances the difference. Given that red cones and green cones respond to similar wavelengths of light, this is just as well: if the differences in signal were not enhanced, we would never be able to distinguish all the colours that lie between red and green. The LGN processes blue-cone signals differently, comparing blue-cone responses to the *combined* responses of red and green cones. What is not blue is red-and-green; in other words, yellow.

If this scheme sounds oddly familiar, look no further than Ewald Hering's theory of opponent colours. Proposed in 1872, this theory holds that two colour pairs – red-and-green and blue-and-yellow – provide the axes for our two dimensions of colour vision. So, was Hering right?

In a sense, yes: there is such a thing as colour opponency and there are two channels for colour information. Hering's physiological insight found its foundation in anatomy. His achievement was real. The trouble is, as experiments grow in sophistication – so that we experiment with very many precisely graduated colours and not just the bold 'primaries' – the muddier our results become. Most worryingly of all, our reported experiences of colour do not tally with the activities of our colour channels as measured by instruments.

This may sound like an anomoly apparent only to a handful of technicians but, given the right conditions, the matching problem expands to threaten the very foundations of colour science. In his essay 'The Case of the Colourblind Painter', the neurologist Oliver Sacks described the sad case of an artist, poisoned by carbon monoxide and involved, hours later, in a car crash. Returning to his studio, his

brightly coloured canvases had become murky grey daubs. The paintings were meaningless and unfamiliar, for there was no colour in them
anywhere. This man, with perfectly functioning eyes, through subtle
but devastating damage to his brain, could no longer see colour.[17]

The London-based neurologist Semir Zeki has considered similar
cases among people and monkeys and has found that, while damage
of this sort robs an individual of the experience of colour, the mechanisms of colour vision continue to function.[18] Asked to match up
coloured counters, people with no experience of colour are still able
match up colours perfectly. They just don't see them.

There is no reason why there should ever be an easy, one-to-one relationship between the structures of our brains and the phenomena we
experience. When, as in the case of opponent colours, this almost
happens, the confusion generated is immense; John Mollon, for one,
believes colour science disappeared into a *cul de sac* for years because
of the general, and mistaken, assumption that anatomy and genetics
had somehow 'discovered' the hiding-place of Hering's opponent
colours.[19]

Because they have no anchor in our experience, our mechanical
models of how the mind works are astonishingly varied, changeable,
and temporary. Inevitably, an analogy is drawn between the mind
and whatever happens to be the most complicated information-
handling system of the time. Up to 300 years ago, the soul was the
fluid part of a complicated plumbing problem, then steam power
provided a better idea of how power and information might be generated and handled in the body. By 1861, electricity was being harnessed
to shuffle information through a telegraph system, and it was generally thought that consciousness was conducted through – surprise,
surprise – a network of electrical switches.

The computer provided the next big, and relatively futile, change
of analogy. In 1966, at the Massachussetts Institute of Technology,
the robotics researcher Marvin Minsky handed one of his graduate
students what he thought would be a relatively easy summer project:
to build a system of machine vision. This raises a complacent smile
on the faces of today's students: how naïve of Minsky, to imagine
that the human brain is like a computer!

The parade of mechanical analogies troubled Ernst Mach, who was to become one of the modern era's great philosophers of science. Mach understood that we needed a new vocabulary to talk about sensations. 'Bodies do not produce sensations,' he wrote, 'but complexes of elements (complexes of sensations) make up bodies.'[20]

Our bodies are not the same today as they were yesterday. Thought by thought, cell by cell, calorie by calorie, the body plays its part in the flux of things. We talk as if bodies are real and abiding, while sensations are fleeting, but sensations have their origins in the physics which contains us. Colours, sounds, spaces, and times are the ultimate elements, and we their temporary nexus. 'By the recognition of this fact,' Mach assures us, 'many points of physiology and physics assume more distinct and more economical forms, and many spurious problems are disposed of.'

Mach's curious, disembodied way of discussing sensation is catching on, but slowly. It is still depressingly easy to find 'new' accounts of consciousness that talk about the brain as though it were a fleshy computer; and a digital computer at that.

In the next, and final, chapter we shall see where this kind of thinking is leading us – for both good and ill.

TEN

Making eyes to see

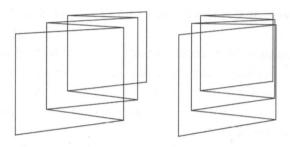

A camera is no more a copy of an eye than the wing of a
bird is a copy of that of an insect. Each is the product of an
independent evolution; and if this has brought the camera
and the eye together, it is not because one has mimicked the
other, but because both have had to meet the same problems,
and have frequently done so in the same way.

George Wald

Around 1790, three men came to Bahrain from what is now the
United Arab Emirates and sought treatment from the American
mission doctor there. He could do nothing for them. Their eyes had
been burned out with a hot iron.

There was, at that time, a story circulating of a European woman
who, while visiting the ladies of the Persian court, came across a
small boy walking about blindfolded. She asked him what the game
was. He replied that he was practising; when he grew up, he was
bound to have his eyes put out sooner or later.

Around 1828, as the Ottoman Empire began its slow decline, it
seemed the practice of blinding one's political enemies fell out of
fashion. But blinding the enemy enjoyed a revival during the First

World War, especially on the Western Front, where mustard gas attacks blinded thousands. Though many recovered some vision, their eyes were irreparably damaged.

Now it is the turn of the New World.

One night in 1994, a year into the United States' involvement in a UN humanitarian mission to Somalia, Emil Bedard and a company of 175 marines found themselves in front of a mob of about 10,000 demonstrators. They were completely defenceless. They had guns – lots of guns – but that was the problem. If they used them, they would kill unarmed civilians, and then the mob would surely kill them.

'Several marines took serious injuries ranging from lacerations to broken jaws,' Bedard recalls. 'As I stood with them, I thought, "How can we deal with this more effectively?"'[1] A year later, he got his answer. In February 1995, US Marine Lt Gen. Anthony Zinni led the 1st Marine Expeditionary Force on a mission to bring out 2,400 UN troops from Mogadishu. For years, he had been lobbying for weapons that would allow his men alternatives to killing people. Of the new weapons he took to Somalia, two hit the headlines. One was a 'disabling' foam so innocuous that evening news broadcasts ran footage of its deployment alongside a clip from the film *Ghostbusters* where the actor Bill Murray is felled by slime.[2] The other was a laser powerful enough to blind people.

The rescue operation was a complete success. The problems the US marines had with their non-lethal weaponry cannot be held against them: they were unfamiliar weapons, deployed for the first time in a dangerous environment. They were used moderately. Commanders on the ground were sufficiently concerned about the risks of permanently blinding civilians they ordered that the lasers be 'de-tuned', turning them into little more than powerful torches.

On 1 March, 1995, commandoes of the US Navy SEAL Team 5 were positioned at the south end of Mogadishu airport when they saw a Somali man aim at them with a rocket-propelled grenade. They used their laser to target him, and a SEAL sniper killed him. The team – trained in non-lethal combat and looking forward to fighting a new, bloodless kind of battle – were not at all pleased: 'We were not allowed to disable these guys because that was considered inhumane,' one complained. 'Putting a bullet in their head is somehow more humane?'[3]

All weapons cause horrible injuries, intentionally or not. The problem with dazzling lasers is that the specific horrifying injury they inflict – blindness – cannot be treated. False legs let you walk, after a fashion; prosthetic arms give you some limited dexterity; but there are no prosthetic eyes. In September 1995, the US Department of Defense banned the development of lasers specifically designed to blind people. In 1996 the ban became global, under the Geneva Convention.

Nonetheless, attempts to produce an 'eye-safe' dazzling laser continue. The Veiling Glare Laser, under development at the Joint Non-Lethal Weapons Directorate in Quantico, Virginia, uses near-ultraviolet light to make the lens of the eye fluoresce; it is painlessly, but effectively, dazzling. The long-term risks to vision have yet to be assessed.[4] Meanwhile, cruder devices find favour in the field. In May 2006, the US military – wary of shooting civilians who fail to heed their warnings at checkpoints – fitted some of their M4 rifles with a laser that is 'like shining a big light in your eyes', according to Lt Col Barry Venable, a Pentagon spokesman. 'Calling them weapons would be a misnomer,' he adds, with an eye to international accords and trigger-happy journalists. 'It's not the "death ray".'[5]

But the use of strong lights to control civilians raises an important wider point – well-made by Emil Bedard, though he is now a vocal and active convert to non-lethal weaponry: '. . . not everyone will accept these capabilities solely because they are intended to be non-lethal. Some will argue that these capabilities, if misused or in the wrong hands, may have catastrophic results.'

Throughout human history, we have sought to take control of vision. The earliest recorded attack on enemy vision dates from the Peloponnesian War, in the 5th century BC, when Spartan forces lit a mixture of wood, pitch, and sulphur under the walls of a city to incapacitate the defenders. For most of this chapter, I want to consider the happier consequences that follow from that desire: our attempts to restore sight to the blind and, more recently, our attempts to build eyes so that machines can be given the gift of sight.

The first part considers the early history of eye medicine, and the difficulties confronting surgeons who restore sight to a person who has had no previous experience of vision. The chief, and besetting, problem with such an operation is that vision is learned through movement. A person with restored sight may be aware of light but they will probably not be able to see. In the second part of this chapter, we will see how the relationship between movement and vision inspired some important physiological studies, and informed the most productive of today's assays at machine vision. Finally, we look at how machine vision and eye medicine are married in the latest developments in prosthetic vision. The devices promise renewed sight to the blind, an enriched visual experience for us all, and a possible future more nightmarish than any a non-lethal weapon could inspire.

1 – 'Suppose a man born blind . . .'

Even his fiercest critic admitted Henry Blackbourne 'could couch cataract well'.[6] Operating in the early seventeenth century, Blackbourne was an itinerant oculist. Oculists were self-appointed surgeons, whose range of operations typically included cures for deafness and the remedial treatment of hare lip and cleft palate. Their signature operation – the one that had them setting up scaffolds on the village green and ringing bells to gather an audience – was 'couching'. A lens afflicted by cataract blocks the view: if the lens is moved out the way of the pupil, the afflicted eye gains unimpeded (if blurry) vision. Since extraction of the lens was not possible at this time without a great deal of trauma, couching needles were used to push the cataracted lens down inside the cushion of the iris.

The operation blinded as many people as it cured, but that was neither here nor there, since cataract, left untreated, is a one-way ticket to total blindness. Blackbourne's critic, Richard Banister, writing in 1622, was more concerned with the way Blackbourne 'would cozen and deceive men of great sums of money by taking incurable diseases in hand.' Blackbourne's greatest weakness was women: 'He was amorously given to several . . . so that his cozening made him fearfully to flee from place to place.'

Banister wanted to get rid of itinerants like Blackbourne. Even more, he wanted to get rid of the 'boundless boldness of many women, who for lack of learning cannot be acquainted with the theoretic part, and yet dare venture on the practical. I believe, scarce three of thirteen hundred can define or describe the names and natures of the hundred and thirteen diseases of the eye . . .'

He had a steep hill to climb. Just as much as Blackbourne, he was working from folklore and guesswork. To give us some measure of how little was known about the eye at this period, we need only look at the records of Walter Bayley. Bayley was Queen Elizabeth I's physician, a learned man who wrote on the preservation of eyesight. Bathing sore eyes in urine is excellent advice – urine is sterile and mildly antibiotic – but in an age that regularly used urine as a facial cleanser, it was hardly ground-breaking. His other big recommendation – drinking ale to strengthen the sight – is even harder to justify.

The problem is more than a mere lack of knowledge. There was a palpable reluctance on the part of the medical establishment to treat the eye with any degree of seriousness. As late as 1666, nearly half a century after Banister's sterling efforts at gentrification, Robert Turner (author of *The Compleat Bone Setter*) gave as treatment for squint: '. . . the blood of a turtle or head of a bat, powdered of course. Or for weak sight, the eyes of a cow hung around the neck.'

In Britain, major advances in eye surgery had to wait until the establishment of Moorfields Hospital, the world's first dedicated eye hospital, in 1805. But there were some glimmers of hope. In 1728, William Cheselden, of St Thomas' Hospital in London, removed cataracted lenses from the eyes of a boy of thirteen. This was not the first such operation to restore sight to a blind person; the earliest we know about gave vision to a thirty-year old man, in 1020. It is, however, the first full account we have of the operation and its aftermath, and it began a modest but growing fashion for making the blind to see.

Studying the case, the French physician Julien Offray de La Mettrie (1709–1851) – author of *Man a Machine* and a notorious materialist – noted the difficulty the boy had in seeing with his newly clear eyes. He had vision but he lacked education in the ways of seeing. The idea that learning has a role to play in sensation reaches forward to our own day, to the work of Gerald Westheimer, Michael Land,

and others, who are studying how the eye learns to sort salient information from the world. It also reaches back to the work of William Molyneux and the empiricist philosopher John Locke, whose dialogue on sensory learning provided Locke's *Essay Concerning Human Understanding* of 1690 with one of its most celebrated passages. Molyneux wrote to Locke:

> Suppose a man born blind and now adult, and taught by his touch to distinguish between a cube and a sphere of the same metal. Suppose then the cube and sphere were placed on a table, and the blind man made to see: query, whether by his sight, before he touched them, could he distinguish and tell which was the globe and which the cube? . . . The acute and judicious proposer answers: not. For though he has obtained the experience of how the globe, how the cube, affects his touch, yet he has not yet attained the experience that what affects his touch so or so, must affect his sight, so or so . . .[7]

Cheselden's operation (and subsequent, improved attempts) seemed to bear this out: curing a person of blindness was one thing; enabling them to see was quite another.

In 1777, Franz Anton Mesmer (1734–1815) ran into dreadful trouble when he attempted to restore the sight of Maria, daughter of Empress Maria Theresa's Court Councillor, Joseph Anton von Paradis. How Maria von Paradis lost her sight – it faded steadily between two and five years old – is not known. It is generally assumed that her blindness was 'psychogenic'; that there was nothing wrong with her eyes, her optic nerves, or the visual centres of her brain, and that her blindness was the product of a dissociative disorder. It was (to put it unkindly) all in her mind. This would certainly serve to explain Mesmer's success. Today, his treatments, which involved the judicious application of what he called 'animal magnetism', carry an unmistakable whiff of snake oil. But the evidence suggests that Mesmer was sincere. Mesmer's life's work was to put the phenomenon of faith healing on a scientific basis. If he failed, it was an interesting failure, revealing, to later researchers, the power of medical suggestion (the 'placebo effect').

When Mesmer met her, Maria was eighteen years old, a musical prodigy (Mesmer's friend Mozart wrote his piano concerto in B flat for her) and a favourite of the Empress. In exerting (at the very least) the power of suggestion, Mesmer was able to unlock Maria's psychogenic blindness, and for a while, she appeared to be recovering her sight. There are many versions of what went wrong and each is more purple than the last. Because Maria's family received a pension because of her blindness, one version has it that her mother forcibly removed Maria from Mesmer's care, slapping her into submission and a return of her blindness. If her mother were truly vicious, this might, for the prurient, suggest a cause: psychogenic blindness is often triggered by traumatic and unspeakable events. Another has it that Mesmer seduced his eighteen-year-old client. I wouldn't put it past him: Mesmer's biography, a bewildering mix of genuine intellectual labour and blatant charlatanism, inspires wry admiration more than moral confidence. However, this story is no more substantiated than the first; we had better dismiss them both.

Much more, revealing, I think, is the hostility Mesmer faced from the doctors who had been treating Maria unsuccessfully for the previous ten years. Their complaints, however much they were motivated by jealousy, were public documents, soberly expressed, and devastating in their simplicity: if Maria could see, as Mesmer claimed, why could she not be taught to identify anything? Why could she not call anything by name? Mesmer had given Maria back her sight, or the suggestion of sight, but Maria did not seem to know what to do with it. Eventually, amid accusations and counter-accusations, the Empress had a quiet word: 'all this nonsense' had to stop. Mesmer, disappointed, left Vienna for Paris, and Maria reacquired her blindness.

It served her well. Able to focus again on the sound world she had inhabited since she was five, Maria became a celebrated composer and musician. She wrote at least five operas, as well as cantatas, *lieder*, and many piano works; her longest European tour lasted three years. I find her loss of vision sad, but I labour under a sighted person's prejudices. I doubt any blind readers will find much to mourn in the life of one of the most accomplished, independent, well-travelled, and successful women of her age.

Mesmer's notoriety, the jaundiced hindsight we apply to his works, and all the frills and furbelows of court gossip and newspaper prattle have obscured a valuable, and utterly typical, case. Maria, restored to the miraculous world of vision, reckoned she was better off blind. As his treatments took effect, Mesmer reports Maria saying: 'How comes it that I now find myself less happy than before? Everything that I see causes me a disagreeable emotion. Oh, I was much more at ease in my blindness.' Richard Gregory and Jean Wallace's moving report from 1963, 'Recovery from early blindness', contains strong parallels with Maria's story.

The two Cambridge researchers tell the story of S B, a man born in 1906, blind since he was ten months old. His sight was restored, by corneal grafts, when he was fifty-two. Gregory and Wallace documented his recovering vision, his attempts to develop a visual memory, and the painful steps he took to relate the world he knew to the confusing play of colours and forms in his newly-opened eyes. They also record his depressions, and how he changed from an ebullient and assertive blind man to an under-confident and morose sighted one. 'His story is in some ways tragic,' they wrote. 'He suffered one of the greatest handicaps, and yet he lived with energy and enthusiasm. When his handicap was apparently swept away, as by a miracle, he lost his peace and his self-respect.'[8]

S B had been trained as a cobbler. Early in their study, Gregory and Wallace thought it would be a pleasant to take him to the Science Museum in London. There, in a glass case, stood a screw-cutting lathe; a machine S B had been familiar with throughout his life, and always wished that he could use.

He was quite unable to say anything about it, except that he thought the nearest part was a handle . . . He complained that he could not see the cutting edge, or the metal being worked, or anything else about it, and appeared rather agitated. We then asked a Museum Attendant for the case to be opened, and S B was allowed to touch the lathe. The result was startling; he ran his hands deftly over the machine, touching first the transverse feed handle and confidently naming it as a handle, and then on to the saddle, the bed and the head-stock of the lathe. He ran his hands eagerly over the lathe, with his eyes tight shut. Then he stood back a little and opened his eyes and said: 'Now that I've felt it I can see.'

We are physical beings. As infants, we map objects by touch, and by relating touch to visual experience, we refine our personal rules about how the visual world is to be interpreted. After years of disuse, we can hardly expect a blind person's brain to be able to disentangle a world full of colour, shade, depth, and movement, nor should we assume that laggard sight can ever 'catch up' enough to be fully integrated with other, mature, experienced senses.

One experiment – disturbing, though necessary if we ever want to treat blindness properly – deprived kittens of visual stimulation at critical moments in their development. The animals never learned to see properly. What is true of that handful of unfortunate cats is just as true of human beings. To date, no person, blind during childhood and restored to sight, has managed to enjoy all the pleasures of the visual world. All of them retain and rely upon at least some 'blind' skills.

2 – How bumping into things teaches us to see

A popular Victorian visual novelty was a handsomely furnished but doll-sized room, nailed over a peephole in a door. One can imagine the consternation of observers, opening the door to this sumptuous interior to be confronted by a different room altogether.

In 1952, Adelbert Ames Jr (1880–1955) took the novelty room to new levels of philosophical sophistication in a series of witty experiments conducted at the Dartmouth Eye Institute, of which he was a founder. Ames had been a painter, until his studies of optics got the better of him. While becoming the world's leading expert on disordered binocular vision, he took considerable delight in fashioning optical illusions to demonstrate the level of guesswork involved in seeing. His most famous construction is the Ames Room, a distorted cube which, looked at through a window, appears regular. However, when two people stand either side of the room, one appears dwarfed, while the other appears to have swelled to giant proportions; their head practically touches the ceiling.

Since sudden changes in size never happen except in *Wonderland*, you would be forgiven for thinking that the illusion must break down at this point. But a photograph taken through the window of an Ames Room (see last page of colour section) shows how persistent the illusion can be. The eye is faced with two irreconcilable absurdities: either the people have shrunk and expanded, or everything else is lined up to fool the eye. As far as the eye is concerned, it is easier to believe that something has gone wrong with the people than that something has gone wrong with everything else.

This is interesting, because it reveals how devoted our eyes are to the laws of optics. Even more intriguing is what happens if you throw tennis balls through the window, or poke at the interior of the room with a long stick. Physically testing the environment reveals its true distorted shape; after a little exploration of this kind, the eye ceases to be fooled. Gradually, it swaps one absurdity for the other, and sees the room for the strange, topsy-turvy space that it is.

Experimental psychologists possess a bottomless appetite for these sorts of games.[9] Hermann von Helmholtz, a keen hiker, once wrote about the effect of bending over and looking through his legs at the surrounding mountains. (Sudden inversion of the view accentuates one's depth perception wonderfully. I tried this minutes before tackling some desperate, broken ascent on the Pic du Midi in the Pyrenees, and given the effect on my morale, I rather wish I hadn't.) In the 1890s, George Stratton of the University of California went one better, and built volunteers several lens and mirror systems to view things from unfamiliar angles. For one experiment, Stratton wore inverting goggles, turning his world upside down; he found that things righted themselves after a couple of days. Once he took the goggles off, it took about the same amount of time for his vision to right itself again.

Did Stratton 'get used' to his inverted vision, or did the world 'spring upright' for him? Are these simply two ways of describing the same experience? This question – which steers perilously close to 'is my green the same as your green' territory – inspired Ivo Kohler, at the University of Innsbruck, to repeat Stratton's work. As far as I can tell, Kohler holds the world record for the longest continuous psychological experiment. For 124 days, Kohler wore glasses fitted with binocular prisms, reversing his field of view left to right. There can be no doubt that he adapted to his strange glasses: he wore them even while riding around Innsbruck on his motorbike.[10] In another experiment, using spectacles that turned the view upside down, Kohler noted that visual adjustment is not a steady process. In the early stages, when the world still appears upside down, touching objects tends to make them look suddenly normal. Even touching things at one remove – for example with a long stick – is enough to turn the world the right way up. In addition, objects right themselves if they seem logically absurd while inverted. A candle might appear inverted until it is lit, when it flips up the right way.

'The different sense-perceptions,' Stratton wrote, 'are organized into one harmonious spatial system. The harmony is found to consist in having our experience meet our expectations.' We do not learn to see through our eyes alone, but by the movements of our bodies through our environment. Learning to see is as much about moving and touching as it is about focusing and converging the eyes. As long as our environment remains reasonably stable, so that objects remain solid, retain their colouring, and don't deliberately line up to appear joined when they're not, we rarely run into difficulty.

Trust the world to teach your senses: this lesson has not been lost on the robotics community. At the Massachusetts Institute of Technology, Rodney Brooks has spent twenty-five years directing an artificial intelligence (AI) laboratory unlike any other. For much of the 1980s, it resembled a grown-up version of the family living room when he was a child of twelve, building robots that could respond to light, wander around the floor, and negotiate obstacles. (His father was a technician in a defence lab and used to bring home the bits.)

Brooks's eighties robots looked like insects, but they are a lot simpler. They have no memory to speak of and little scope for action. What they do have, in spades, is input. They are equipped to sense every bump, knock, and scrape. Brooks's team did not teach their robots anything. They did not programme them, in the usual sense. They did not give them any *a priori* information about the environment. They trusted that the world itself, perceived through primitive senses, would shape their robots' behaviour. The lab turned out the most life-like robots of their time, robots that developed simple responses to simple hazards. Brooks's approach – dubbed 'embodied' or 'situated' AI – revolutionised artificial intelligence.

In 1997, Brooks featured in a bizarre documentary by Errol Morris. *Fast, Cheap and Out of Control*, about people whose work drew its inspiration from the animal world. Its title very neatly sums up the kind of world Brooks's robots are ushering in. Why build one heavy, slow, extremely expensive and complicated machine to do a task, when you can share that task out among a swarm of simpler, smaller, dumber machines? Why build a single, impossibly expensive robot to map Mars, when you could more easily cloud the Martian skies with tens or hundreds of thousands of tiny, expendable, mechanical 'bees'?

Fast, cheap, out-of-control robots. They have no complicated instructions. They have no sense of mission. What they have is senses; you could say that they are little more than 'senses on legs'. As, year by year, their senses are made subtler, so they are able to develop a more refined response to their environment.

Robots built with Mars in mind are the first recipients of a new generation of mechanical eyes; organs which mimic the navigational abilities of compound eyes.[11] Like dragonflies they detect the line of the horizon, using ocelli made up of a handful of ommatidia. Like bees, they measure their airspeed and their nearness to obstacles by optic flow, noting the rate at which textures pass neighbouring ommatidia. Like ants, they navigate using polarised light. The latest really good prototype (it can hover over a target using its own senses, and needs no instructions) is a miniature helicopter built by the Australian researchers, Javaan Chahl and Mandyam Srinivasan. With a length of 1.5 metres and a weight of seven kilogrammes, it is much too big to go to Mars – not so much a bee as an albatross – but it represents a huge logistical advance over poor Sojourner Rover. This, the Mars Pathfinder's mobile unit, crept off its ramp on 5 July, 1997. It had no senses and had to be directed centimetre by centimetre, across the Martian surface, from Earth, 120 million miles away. In thirty days, it covered fifty-two metres.

*   *   *

Brooks has not been slow to capitalise on the practical applications of his work. (His CV as an entrepreneur is as long as his list of academic achievements.) We should not, however, lose sight of why robots like these are made: they are built as experiments, to answer questions about the nature of intelligence. The MIT lab's headline achievement of the 1990s was Cog, a humanoid robot with moving hands, arms and eyes. It had a sense of touch, and a sense of vision. It could reach out and touch objects, and it could both see and feel what it touched.

Cog has gone into honourable retirement now. With its experience to guide them, what will Cog's successors make of their world? Will they go a step further, and take an interest in things? Will they learn to point? Will they learn to follow each other's gaze, or even – wonder of wonders – acquire theory of mind? If being human is something we learn through copying the visual behaviours of our parents and our peers – as the work discussed in Chapter Five suggests – any serious attempt to build a thinking machine will have seriously to consider not just its physical senses, but also its social environment.

This idea dates back surprisingly far, to the work of the Russian Lev Vygotsky, who put forward this view in the 1920s. Unfortunately, his work was suppressed by the Soviet authorities, and was not available in the West until the 1960s.[12] 'The mechanism of social behavior and the mechanism of consciousness are the same,' Vygotsky wrote. 'We are aware of ourselves in that we are aware of others; and in an analogous manner, we are aware of others because in our relationship to ourselves we are the same as others in their relationship to us.'[13]

One Cog may never be enough. We may have to build a whole society of Cogs; a task so patently Herculean that some critics wonder whether there is much future in robotic approaches to AI.[14]

3 – New eyes for old

We do not know his name – anonymity has been preserved for his family's sake – and his case has never been discussed in any formal scientific papers. We know he was blind and that, towards the end of the 1990s, he volunteered for a trial of an innovative system of prosthetic vision, using brain implants. We know he found the implants a nuisance because one day, not really thinking about it, he took hold of the wires connecting his brain to the machinery, and tugged.[15] His death, following a massive infection, brought to a crashing halt vision science's most extraordinary adventure to date: the attempt to bypass the eyes entirely, and bring sight to people who are profoundly blind. But the project is far from abandoned, and how such a technology is applied will become an important ethical debate for the next generation.

People blind from birth or early childhood have difficulty learning to see. However, in the developed world, the vast majority of blind people lose their sight in adulthood. They know how to operate in the visual world, and would certainly benefit from having their sight restored. Because there are so many different ways to lose sight – so many things that can go wrong in the complex machinery of vision – the pressure is on to develop a system that can address all causes of blindness at once: a final solution to the problem of sightlessness.

Work on projecting pictures directly into the brain began in Cambridge, England, in the early 1960s, under the leadership of Giles Brindley. Although there was no realistic prospect that these experiments would restore her sight, a volunteer – a fifty-two-year-old woman who had been blinded by glaucoma – agreed to the installation of eighty electrodes over her visual cortex and, under her scalp, an array of radio receivers. The idea was that a transmission, picked up by a particular receiver, would stimulate a corresponding electrode, and the tiny charge emitted by the electrode would stimulate a portion of the woman's visual cortex. But what would she see? Brindley hoped to exploit a finding by the pioneering Swedish neurologist Salomon Henschen. In 1893, Henschen had been studying patients who had sustained injuries to their visual cortex. Signals which start their journey in the retina arrive at the area of the visual cortex which today we call 'V1.' Henschen tried matching the location of wounds to V1 to the damaged areas of his patient's visual field, and met with a success that was positively daunting: he found that the scene captured by the eyes is literally mapped over the surface of V1.

It is, admittedly, a funny sort of map: upside-down, yes, and also back-to-front, inside-out, sides-to-middle, convoluted and every other form of twisted you could name. Nonetheless, his discovery –

that there really is a screen (of sorts) in the brain – was confirmed during the Russo-Japanese War and during World War I in studies of wounded soldiers.

By installing eighty electrodes over his volunteer's V1 region, Brindley hoped that he had taken the first step in transmitting intelligible maps onto a blind person's visual cortex.

In theory, Brindley's system might one day be used to provide his subject with a kind of crude, eighty-pixel system of prosthetic vision. But neither Brindley nor his volunteer expected that day to come any time soon.

It didn't: predictably, less than half the woman's electrodes generated any kind of visual image. The others caused her to see stars. And while some electrode responses caused her to see a single star in a single location, others made her see several stars at once. There was another problem: matching the location of the electrodes to the location of the stars proved problematic. According to Henschen, the correspondence should have been exact. But trying to replicate the rippled surface of V1 with a rigid array of electrodes and receivers proved heinously difficult; further complicated by the fact that, whenever the volunteer moved her eyes, the stars appeared in different parts of her visual field.

None the less, Brindley and his collaborator, the neurosurgeon Walpole Lewin, had every reason to be optimistic. In 1968 they wrote:

> Our findings strongly suggest that it will be possible, by improving our prototype, to make a prosthesis that will permit blind patients not only to avoid obstacles when walking, but to read print or handwriting, perhaps at speeds comparable with those habitual among sighted people.[16]

Another major player in the field was the American prosthetist, William Dobelle (1941–2004). Dobelle's life-long interest in what he called 'the spare parts business' began at home in Massachusetts; he was just thirteen when he applied for his first patent for an improved artificial hip. The senses were his particular fascination; lectures delivered in the mid-Seventies predicted just about every major advance and approach in sensory prosthesis for the next thirty years.[17]

Like Brindley in the UK, Dobelle drove forward primary research with a series of ground-breaking operations. In 1978, he and his colleagues implanted an array of sixty-four electrodes in the brain of a genial, thirty-five-year-old Brooklynite, 'Jerry', whose optic nerves had been severed by a gunshot wound. Though their system was in essence the same as Brindley's, Dobelle's team got better results, not least because the devices used in such operations were growing smaller, year after year. Dobelle's study was reassuring: V1 did 'map' the visual scene, as Henschen had claimed. But there were still problems: like Brindley's, Dobelle's team found that many electrodes produced multiple stars. More seriously, they found that their equipment, if used continuously, dangerously overheated the brain.

Both Brindley and Dobelle reined in their expectations. With luck, miniaturisation would eventually solve the heating issue. Meanwhile, their systems of induced 'dot vision' seemed best suited, not to vision, but to the reading of Braille. Getting blind people to see a blind script sounds like a waste of time, but the results from both teams were impressive. By 1979, Dobelle's patients could read Braille by 'sight' five times more quickly than they could by using their fingers.

The wait went on. And on: for nearly twenty years, no major human experiment was conducted. What would be the point? Until the stimulating electrodes were reduced to the size of individual neurons, the prospects for this sort of prosthetic vision were stalled. In 1996, a team led by E M Schmidt implanted the first such microelectrodes into the cortex of their volunteer, a forty-two-year-old woman, blind since the age of twenty. The results were excellent, and other human trials quickly followed. In hindsight, perhaps too little thought was given to the greater risks of using this new equipment. The microelectrodes were so fine, they could be implanted directly into the brain, but their low output meant that they had to be physically connected to the system's machinery. This meant wires coming out from the brain, vastly increasing the risk of infection and accident. Ultimately, for one volunteer at the US National Institute of Health, it meant his death, and an immediate moratorium for the project.

For both Brindley and Dobelle, who had waited so long to see their visions realised, the death marked a sea-change in their careers. Brindley began his professional career in neuroscience (and had supervised through his doctorate the young David Marr, who went on to be a hugely influential figure in the study of visual perception). By the eighties, his interest in prosthetics had led him into many diverse areas, from spinal injury to urology. He became world-famous in 1983, by dropping his trousers at an American Urological Association meeting in Las Vegas. It was a performance worthy of a Mel Brooks film: 'he walked down the aisle and let us touch it', one attendee remembered; 'People couldn't believe it wasn't an implant.' Brindley was demonstrating how a powerful muscle relaxant, injected directly into the penis, can generate an erection. It is generally reckoned that Brindley's show-stopping lecture ushered in the Viagra age.

Dobelle, on the other hand, who had once enjoyed 'spare parts' almost for their own sake – he made significant contributions to hearing implants and artificial hearts – had been deeply bitten by the vision bug. Frustrated by what he saw as the needless new restrictions on his research, he moved to Portugal and continued his work with human volunteers there. It was a decision that led him away from the scientific mainstream; he no longer published in the scientific journals, preferring to talk to the press. He had plenty to talk about: by 2000, he had given restricted vision back to Jerry, his volunteer of 1978.

The 'Dobelle eye' Jerry demonstrated to astounded journalists consists of an ultrasound scanner, mounted on spectacles, which feeds spatial information to a 10lb computer worn at the waist. (Advances in computers were crucial to the system's success: only a year before, Jerry's apparatus had weighed two hundred times that and was as big as a bookshelf.) Once the information has been simplified, to provide information only about the edges of objects, it is broadcast to the sixty-four platinum electrodes implanted on the surface of Jerry's visual cortex. Jerry sees enough of this information to be able to read two-inch tall letters at a distance of five feet, and navigate hitherto unexplored sectors of the New York subway system.

Dobelle implanted arrays in seven new volunteers, and announced the development of a 512-electrode system which, in mass-production, would cost no more than the training bill for an average guide dog. It was a project he was still working on at his death, aged 62, in 2004.

*   *   *

The future of systems like the Dobelle eye is in doubt. They began life as spin-offs from studies into how the brain handles electrical stimulation. Now that the basic research is done, stimulating the brain to manufacture eyeless vision seems risky, long-winded, and even unnecessary.[18]

A major selling point for such systems has been that a retina stricken with blindness quickly deteriorates. A system of prosthetic vision has to stimulate the brain, because following the loss of sight, all parts of the retina lose the ability to see. However, this is not the case. Although common blinding conditions like macular degeneration destroy the architecture of the retina, its nerve cells remain remarkably healthy and functional. Even when the photoreceptors are destroyed, as happens in retinitis pigmentosa, the ganglion cells to which they connect stay in a workable condition for decades.[19] If there were a way of installing a plate of photosensitive 'cells' in the eye, could we get it to talk to the undamaged neural layers of the retina?

Until very recently, artificial retinas were a completely unrealisable fantasy; only since about 2004 have we been able to contemplate building something so tiny and so finely worked. Now that they have become a real possibility, vast sums are being invested in their development.

But where should they go? The back-to-front arrangement of the vertebrate eye makes it difficult to know where to put a prosthesis. Should it sit on top of the retina, covering the ganglion cells but far away from the site of the old, dysfunctional photoreceptors? Or should it go under the retina – with all the risks and dangers such a procedure implies – so that it can take over the job of the dead photoreceptors? Sod's Law dictates that whichever approach I pick as the most promising, it will have been abandoned by the time you read these words. Still, in spite of all the risks, and some recent, rather worrying data on how quickly they corrode, putting a prosthesis under the retina seems the approach most likely to succeed[20]. A prosthesis of this sort only needs to detect light; visual processing can be left to the surviving layers of the retina. A prosthesis that sits on top of the retina and talks directly to ganglion cells has to do all this processing on its own – a daunting task.

Two teams – one from America and one from Germany – lead

the race to develop a working sub-retinal prosthesis. Their contrasting methods and approaches reflect the cultural differences in the two scientific communities. The German project is funded by the German Federal Research Ministry, as part of a multi-million-euro project to realise an 'intelligent retina'. It has yet to proceed to human trials. The American team is a company called Optobionics, founded by Dr Alan Chow, a paediatric ophthalmologist from Illinois, and his brother Vincent, an electrical engineer. After eight years of animal trials, the Chows announced that their patented prosthetic retinas worked well enough that the animals' brains were responding to their signals. They subsequently found that they were wrong; they had not taken into account the possibility that the animals' visual centres were responding to infra-red radiation.

Nonetheless, the US Food and Drug Administration approved their application to proceed with clinical trials. To date, ten volunteers have been implanted with the devices, and they all report improvements in their vision. However, all of them had at least a little residual vision, which makes assessing small improvements very difficult. What is more, they all reported improvements across the whole of their visual field – a neat trick, given that the Chows' arrays are implanted only in the periphery of their vision. The improvements cannot be because the devices work the way the Chows thought they would. At best, the prosthesis is 'encouraging' the rest of the retina to rewire itself. If true, this would be truly wonderful but it seems more likely that the volunteers are simply reporting the kind of small visual changes that you would expect after major ocular surgery.

Assume it happens. There seems no very good reason why it won't. Artificial retinas will not only renew, but may even eventually extend, the capacities of our eyes.

The eye is tuned, by evolution, to our prehistoric needs, but we have since utterly transformed ourselves and our world. By building our own retinas, we can custom-build a different, broader visual experience for ourselves, and today's remedial surgery for blindness will very soon be extended to elective surgery for better, broader sight. Artificial retinas can easily be tuned to see other parts of the electromagnetic spectrum. If a blind person can see with a couple of retinal chips, why not adapt one chip for daylight use and one for night? And if blind people can be equipped for night vision, why shouldn't sighted people get in on the act?

Come to think of it, why stop at light? An artificial retina might conceivably be sensitive to a whole host of novel sensations. Anyone for radar? Spectroscopic analysis, anybody?

The Information Revolution is looking a little tired. The Perception Revolution, meanwhile is cracking its way out of the egg.

4 – What you see is what you get

Flowers, crystals, busts, vases, instruments of various kinds, &c., might thus be represented so as not to be distinguished by sight from the real objects themselves.

Charles Wheatstone[21]

Imagine you are a surgeon, conducting keyhole surgery.

Not the easiest job in the world: first, where are you supposed to look? At your hands? At the point where your instruments disappear into the tiny incision? At the ultrasound monitor, mounted above the table, which shows what you're up to inside your patient? Frankly, you're all over the place. Your attention is scattered and diffused. No wonder the trustees want to replace you with a robot.

But wait – put this pair of goggles on. The goggles have ordinary clear lenses, and a casual glance will not reveal their secrets. But you know that inside these goggles there is a microwave receiver, which tracks the motions of your head as you work. That information is sent, via a cable or a radio link, to a computer, which projects images on to the partly-silvered surface of your goggles. It's a kind of head-up display, of a very special kind. Because the computer knows where you are looking, it can paste images on to your goggles in such a way that they blend seamlessly with your surroundings.

Now, when you operate, it appears that you're actually looking inside the patient's body. Hey presto: you have X-ray eyes.

*   *   *

Funny how some stories sound futuristic, even when they are long out of date. I wrote that story for *Wired* magazine around 1996. Since then, visual prosthesis has advanced almost beyond recognition and with it has come a huge improvement in our ability to sow our individual visual worlds with virtual, pre-processed information (a process dubbed 'augmented reality' or 'AR'). News stories about simple head-up displays and augmented-reality goggles share column inches with tales of lasers that map images across the retina as though it were a television screen, induction loops that excite the optic nerve, and trans-cranial magnetic stimulators that conjure visual fancies in the brain.

Some of these stories are more fantastic than others, but even the most speculative eventually acquire a grain of truth, as the technology seems hell-bent on fulfilling every tech journalist's moistest wet dream. Each year, as half a million children go blind for want of a bit of butter, a telephone directory's-worth of articles appear, and they all begin the same way: 'Imagine you are a fighter pilot.' 'Imagine you are an astronaut.' 'Imagine you are a deep-sea diver . . .'

Here's an idea: give everyone a pair of goggles. With them, and access to a gargantuan shared image database (are the good people at Google listening?) the everyday might become a canvas for us all. We might paste *Gormenghast* over Pall Mall, or glimpse Dickens's London down Fleet Street. Driving to Edinburgh for work, we might choose to give it the skyline Stevenson knew and loved. Goggles on, and every day becomes an adventure in new territory. We might go shopping in the drowned London of Wyndham's *The Kraken Wakes*, catch a plane to *Blade Runner*'s LA, make a Narnia of Birmingham, an Oz of the New Forest, or erect Middle Earth on Salisbury Plain . . .

Map fantasy over reality! Free the world of physical constraints! Build Lang's Metropolis on Earth!

Those of us old enough to remember a world without the Internet will recall that we have heard this kind of talk before. Indeed, in the early twentieth century, the self-same millennial promises were made about another innovative medium: cinema. Will augmenting our individual visual worlds free us of the constraints of space, time, and the body?

I wonder.

One of the more frightening things I stumbled across while researching this book was the illustration accompanying an article about Augmented Reality in the April 2002 edition of *Scientific American*. The accompanying illustration took the author Steven Feiner's lead in suggesting a future in which virtual images are mapped contextually over the outdoor environment. The images were captions, laid across an observer's view of a city street:

Near a bus stop: 'Local #23 NEXT BUS 30 Seconds.'
Over an office building: 'OFFICE SPACE AVAILABLE Contact Megalopolis Realty cityspace@universal.tech'
Outside a cinema: 'FEATURES Jaws of Terror 2:00, 4:00, 6:00'

Is this the best we can do? Are our eyes to be stuffed with nothing better than the usual commercial and political blandishments? Will augmented reality revolutionise our way of being, or will it prove, like so much of the Internet, to be just a new way of doing business as usual?

If we are going to supplement the eye, it is high time we considered what we are going to supplement it with. What do we want our new sensory world to look like? What do we want from our eyes?

In 2000, William Dobelle gave Jerry back his sight; enough that he can (just about) navigate a subway system. But Jerry's vision is, by any measure, very poor. The visual world is very complex. The images captured by the 2000-vintage Dobelle eye were excessively simple. This is why Dobelle and his team went to such pains to program the Dobelle eye's computer to simplify the scenes it captures, reducing them to a few bright lines. In the summer of 2000, Dobelle was working with Jerry at his house when Jerry's eight-year-old son, Marty, piped up: 'You guys are out-of-date. Why don't you take digital signals straight from the TV or computer?'

Now Jerry wants to deal stocks and shares over the internet.

He wants to plug his eyes into the New York Stock Exchange.

'He's hot for that', Dobelle said.[22]

EPILOGUE

The invisible gorilla

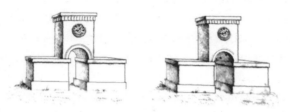

The story of the eye is filled with material wonders: mirror eyes, fibre-optic eyes, and eyes crammed full of exquisite lenses; shells that see and spiders that steer by the sun; eyes less than one-tenth of a millimetre across, and eyes one third of a metre across. Birds can focus on two things at once. Tube worms see with their feeding tentacles, chitons with eyes scattered like pimples over their shells. The fish *Bathylychnops exilis* has four eyes: one set to look up, and one set to look down; but at what? Nobody knows. The surface-feeding fish *Anableps anableps* can see clearly both in and out of the water. Mantis shrimps use polarised light to swap messages that no other animal can see. Even a kilometre below the sea's surface there is, incredibly, still plenty of material my story could have drawn on, if only there were room. Bioluminescent photophores tattoo the sides of lanternfish, fang-toothed fish dangle lures baited with light-making bacteria, and one deep-water fish, *Aristostomias*, communicates with others by generating pulses of red light – a colour no other neighbouring species can see.

The story of the eye is epic: this book is but the shortest précis of its wonders.

*   *   *

Imagine a world without sight.

Colour does not exist. There is no plumage, no camouflage, no courtship display, no phosphorescence in the deep ocean trenches. No flowers.

What a dull world, and what a simple one. Animals develop behaviour to respond to information. Intricate behaviours can never evolve in the absence of enough information. And since behaviours are the bedrock of consciousness, in a world without sight, consciousness will never arise.

Because humans enjoy the luxury of consciousness, we can appreciate seeing for its own sake. For us, seeing isn't necessarily tied to any other useful activity. We tend, naturally enough, to concentrate less on the act of seeing, and more on the view; and assume that the eye is simply a window on the world.

But humans too have their very specific, and idiosyncratic, ways of seeing.

Humans have an exceptionally narrow, acute focus of concentration, coupled with a taste for studying social and transactional behaviour. Of course, what we see reflects the real world, but we should never assume our eyes are telling us the whole story. Welcome to the world of John Grimes: a world where birds and swimming-suits change colour, people swap hats and heads, and nobody seems to notice.

Grimes, of the University of Illinois, presented his first work on 'change blindness' in 1992, at a conference on perception in Vancouver, British Columbia. Sitting in the audience was the American philosopher Daniel Dennett. Only the year before – in a book with the deliberately provoking title *Consciousness Explained* – Dennett had attacked the idea that we carry around pictures of the world inside our brains, saying it was much more likely that we make assumptions and use our senses to confirm or update these assumptions, as and when necessary. So, our assumptions never feel like assumptions, and our illusion of seeing a wide field of view in detail never feels like an illusion.

Even Dennett was startled by Grimes's findings. Grimes's experiments exploited the fact that when our eyes saccade, we are functionally blind. (If this weren't the case, vision would resemble

hand-held camera-work; all nauseating sweeps and irritating judder.)
Grimes set his volunteers to study photographs on a computer screen.
When their eyes performed a saccade, the computer quickly swapped
one photograph for another. In Grimes's first experiments, the difference in the images was subtle. Hats worn by people in one photograph
might be swapped round in the next.

Nobody noticed.

So Grimes tried swapping heads.

Nobody noticed.

Grimes made objects change colour.

Nobody noticed.

'I wish in retrospect that I'd been more daring, since the effects
are stronger than I claimed', said Dennett, following Grimes's original
presentation.[1] By 1996, Grimes had pushed his study even further,
swapping a red parrot that filled a good quarter of his computer
screen, for a green one. One in five of his volunteers failed to notice.

Rather than remember all the incidental details of the world
around us, we rely on our ability to spot changes. If change happens
when our eyes are saccading, we are blind to the change, and it passes
us by. Happily, very few things in the world change in the time it
takes for our eyes to saccade. But what about when we blink? And
what if something comes between us and what we're looking at,
blocking our view?

A sunny day on the campus of Cornell University, in New York
State, 1998. The passer-by looks fifty-something. His magnificent
whiter-than-white hair is the first thing you notice. Perhaps he is a
professor. In any event, he seems to know his way around; when a
young man stops him to ask for directions, he happily obliges. A
moment later, two men barge between the student and the professor.
They're carrying a door. For a brief moment, the professor and the
young man are separated.

Now the video gets interesting. The camera is looking down on
the scene from a height; from here, we can see what the professor
can't: as the door passes between them, the boy asking for directions
swaps places with one of the men carrying the door.

The interruption passes. The professor carries on giving directions;
the new man listens and nods attentively. He is wearing a different

shirt, a different jacket, and we can safely assume that his voice is different, but the professor doesn't notice. Though the man has changed, the professor did not witness the change, and so he assumes the man is the same as the one he was talking to a moment ago. Why would he think otherwise? You might think his memory would wake him up to what is going on; on the contrary, fully half the unwitting subjects of this devious and very funny experiment – dreamt up by the Harvard psychologist Daniel Simons and his colleague Daniel Levin of Kent State University, Ohio – never spotted the swap.

Memory is expensive. It is far cheaper for us to rely on the present moment for our information. If the world were full of trickster psychologists, we may have had to evolve differently. Happily, it's not, and we're really rather easily tricked. We don't need to be blind to changes to miss them; because we only see two degrees of the world with any clarity, we tend to miss things that happen outside our focus of attention, and the more we attend to something, the more extreme our 'attention blindness' becomes.

This is the thesis of Simons' most celebrated experiment, in collaboration with Harvard's Christopher Chabris.

You are about to watch a short video of two teams of students playing basketball in a corridor. One team wears white, one wears black. You have to decide which team passes the ball most often. Concentrate: the answer's important.

Run tape.

Have you got your answer? Good. Now we're going to watch the video again. This time, don't worry about the number of passes. Just sit back and watch the game. Do you see anything you hadn't spotted before?

That's right.

A gorilla.

A two metre tall pantomime gorilla just wandered through the middle of the basketball game. If you were able to concentrate, and put out of your mind the fairly obvious fact that Simons must have something screwy up his sleeve, there's a fifty per cent chance you missed it – even when it stopped and waved at you.

There's no camera trickery, no movement of images during a

saccade, no sneaky undergraduates barging past you with a door. All Simons did was ask you to follow a game. For fully half the people who see this video, the ball-counting task is absorbing enough to make them miss an entire gorilla.

In 1909, the zoologist Jacob von Uexküll proposed that the significant world of an animal was the sum of things to which it responds. The rest goes by almost unnoticed. And is it not ironic, that in 538 million years of natural selection, eyesight should evolve from a simple light-detecting cell, pass through numerous variations and generate countless different ways of seeing, and come at last to serve as the dominant sense of the planet's dominant species – an animal who sees only what it wants to see?

FURTHER READING

General

Gregory, R.L. 1998. *Eye and Brain: The Psychology of Seeing*. Fifth edition. Oxford, Toyko: Oxford University Press. [Uses optical illusions to reveal the mechanisms of perception.]

Helmholtz, H. v. (1856–67. Trans. J.P.C Southall 1962. *Handbuch der Physiologischen Optic*. New York: Dover. [The foundation stone of modern vision science, still relevant and readable.]

Marr, D. 1982. *Vision*. New York: WH Freeman & Co. [Striking, still-current model of vision, steeped in computer metaphors.]

Morgan, M. 2003. *The Space Between Our Ears: How the Brain Represents Visual Space*. London: Weidenfeld & Nicolson. [A good recent attempt to shed computer-based metaphors when discussing vision and mind.]

Oyster, C.W. 1999. *The Human Eye: Structure and Function*. Sunderland, Mass: Sinauer Assoc. [Sumptuous 800-page anatomy of the eye, with many fascinating digressions into the history, biography and theory of vision.]

Wade, N. 2000. *A Natural History of Vision*. Massachusetts: MIT Press. [Vision theorists – often in their own words – from the Greeks to the Victorians.]

Zajonc, A. 1993. *Catching the Light: the Entwined History of Light And Mind*. Oxford: OUP. [Thoughtful, often surprising history of light and colours;

readers scandalised by my omission of Goethe from these pages had better go here to decompress.]

Prologue – Youth and age

Daw, N.W. 1995. *Visual Development*. New York: Plenum.

Hamer R.D. & Schneck M.E. 1984. 'Spatial summation in dark-adapted human infants' *Vision Research* **24/1**, pp. 77–85.

Weale, R.A. 1992. *The Senescence of Human Vision*. Oxford: OUP.

Wiesel T.N. 1999. 'Early explorations of the development and plasticity of the visual cortex: a personal view.' *Journal of Neurobiology* **41**, pp. 7–9.

One – The commonwealth of the senses

Catania, K.C. 1999. 'A nose that looks like a hand and acts like an eye: the unusual mechanosensory system of the star-nosed mole.' *Journal of Comparative Physiology A: Sensory, Neural, and Behavioral Physiology* **185**, pp. 367–372.

Findlay, J.M. & Gilchrist I.D. *Active Vision*. 2003. New York, Oxford: OUP. [A technical analysis of the relationship between eye movements and perception.]

Frisby, J. 2004. 'Bela Julesz 1928–2003: a personal tribute.' *Perception* **33**, pp. 633–637.

Gilchrist I.D., Brown V, Findlay J.M. 1997. 'Saccades without eye movements.' *Nature* **390**, pp. 130–1.

Julesz, B. 1995. *Dialogues on Perception*. Cambridge, Massachusetts: MIT Press.

Wade N.J. 1994. 'A selective history of the study of visual motion after-effects.' *Perception* **23**, pp. 1111–1134.

Wade, N. & Tatler, B. 2005. *The Moving Tablet of the Eye: the Origins of Modern Eye Movement Research*. Oxford: OUP.

Wade, N.J. 2005. 'The original spin doctors – the meeting of perception and insanity.' *Perception* **34/3**, pp. 253-260. [describes Purkinje's researches into vertigo.]

Walls, G.L. 1962. 'The evolutionary history of eye movements.' *Vision Research* **2**, pp. 69–80.

Weiskrantz, L. 1986. *Blindsight: A Case Study and Implications*. Oxford: Clarendon Press.

Wheatstone, C. 1838. 'Contributions to the physiology of vision 1: on some remarkable, and hitherto unobserved, phenomena of binocular vision.' *Philosophical Transactions of the Royal Society of London* **128**, pp. 371–394.

Yarbus, A.L. 1967. 'Eye movements during perception of complex objects', in L.A. Riggs, ed., *Eye Movements and Vision*. New York: Plenum Press, pp. 171–196

Two – The chemistry of vision

Boll, F. 1877. 'Zur anatomie und physiologie der retina.' *Arch. Anat. Physiol.* (**Physiol. Abt 4/35**). Trans. Hubbard, R. 2003 in *Vision Research* 17, pp. 1249–65.

Campion-Vincent, V. 1999. 'The tell-tale eye.' *Folklore* **110**, pp. 13–24.

Carpenter, K.J. 2003. 'A short history of nutritional science: part 1 (1785–1885).' *Journal of Nutrition* **133**, pp. 638–645.

Evans, A.B. 1993. 'Optograms and fiction: photo in a dead man's eye.' *Science-Fiction Studies* **20/3**, pp. 341–61.

Goulet, A. 2005. 'South Sea daggers and the dead man's eye: foreign invasion in *fin-de-siecle* optogram fiction.' *Cahiers Victoriens et Édouardiens* **61**.

Lehman, S. 2004. 'George Wald: a scientist who studied vision and saw a world in crisis.' *Optics and Photonics News.* **15/2**, p. 18.

McLaren, D.S. *The Control of Xerophthalmia: a Century of Contributions and Lessons*. Basel: Sight and Life. Available at: www.sightandlife.org/booksAll/BooksHTML/allBooks.html

Sommer, A. & West, Jr. K.P. 1996. *Vitamin A Deficiency: Health, Survival, and Vision*. New York, Oxford: Oxford University Press.

Wolf, G. 2001. 'The discovery of the visual function of vitamin A.' *Journal of Nutrition* **131**, pp. 1647–1650.

Wolf, G. 2002. 'The experimental induction of vitamin A deficiency in humans.' *Journal of Nutrition* **132**, pp. 1805–1811.

Wolken, J.J. 1995. *Light Detectors, Photoreceptors, and Imaging Systems in Nature*. New York, Oxford: Oxford University Press.

Three – How are eyes possible?

Axelrod J. 1988. 'An unexpected life in research.' *Annual Review of Pharmacology and Toxicology* **28**, pp. 1–23.

Conway-Morris, S. 1998. *The Crucible of Creation: the Burgess Shale and the Rise of Animals*. Oxford: Oxford University Press.

Conway-Morris, S. 2003. *Life's Solution: Inevitable Humans in a Lonely Universe*. Cambridge: Cambridge University Press, p. 166.

Darwin, C. [1859] 1982. *The Origin of Species*. London: Penguin Books.

Gehring, W.J. 1998. *The Homeobox Story*. New Haven, CT: Yale University Press. p. 207.

Gehring, W.J. 2002. 'The genetic control of eye development and its implications for the evolution of the various eye-types.' *The International Journal of Developmental Biology* **46**, pp. 65–73.

Gehring, W.J. 2002. 'The journey of a biologist: Balzan Prize winner 2002 for Developmental Biology.' Available at www.balzan.com/en/preistraeger/gehring.cfm

Gehring, W.J. 2004. 'Historical perspective on the development and evolution of eyes and photoreceptors.' *The International Journal of Developmental Biology* **48**, pp. 707–717.

Parker, A. 2003. *In the Blink of an Eye*. London: Perseus Books. [Lively and original account of the role of vision in the Cambrian, though Simon Conway-Morris's review, 'On the first day, God said . . .' (*American Scientist* **91/4**) bursts Parker's bubble somewhat.]

Wald, G. 1968. 'The molecular basis of visual excitation.' *Les Prix Nobel en 1967*, ed. Granit, R. Stockholm: Nobel Foundation.

Walter, N. 2002. 'From transdetermination to the homeodomain at atomic resolution – an interview with Walter J. Gehring.' *The International Journal of Developmental Biology*, **46**, pp. 29–38. [Arguably the worst title ever given to a good interview.]

Four – The adaptable eye

Dawkins, R. 1996. *Climbing Mount Improbable*. London: Penguin Books. [Contains a spectacular account of eye evolution. The figure by Michael Land – a whimsical-but-serious map of eye variety – is the cherry on a very rich cake.]

Exner, S. 1891. *Die Physiologie der facettierten Augen von Krebsen und Insekten*. Vienna. Trans. Hardie, R.C. 1989 as *The Physiology of the Compound Eyes of Insects and Crustaceans: a Study*. Berlin, New York: Springer Verlag.

Fortey, R. 2001. *Trilobite: Eyewitness to Evolution*. London: HarperCollins.

Gislén, A., Dacke, M. Kröger, R.H.H. Nilsson, D.E. Warrant, E.J. 2003. 'Superior underwater vision in a human population of sea gypsies.' *Current Biology* **13**, pp. 833–836.

Land, M.F. 1997. 'Visual acuity in insects.' *Annual Review of Entomology* **42**, pp. 147–177.

Land, M.F & Nilsson, D.E. 2002. *Animal Eyes*. New York, Oxford: Oxford University Press. [The best single-volume work on eye variety.]

Nilsson, D-E. 1989. 'Optics and evolution of the compound eye.' *Facets of Vision*. (Stavenga, D.G. & Hardie, R.C. eds). Berlin-Heidelberg: Springer-Verlag. p. 3075.

Wald, G. 1953. 'Eye and camera.' *Scientific American Reader*. New York: Simon & Schuster. pp. 555–68.

Walls, G.L. 1942. *The Vertebrate Eye and its Adaptive Radiation*. Bloomfield Hills, MI: Cranbrook Institute of Science.

Zucker, C.S. 1994. 'On the evolution of eyes: would you like it simple or compound?' *Science* **265**, pp. 742–743.

Five – Seeing and thinking

Byrne, R. & Whiten, A. (eds) 1988. *Machiavellian Intelligence*. New York, Oxford: Oxford University Press, pp. 34–49.

Darwin, C. 1872. *The Expressions of Emotions in Men and Animals*. London: John Murray.

Ekman, P. 2001. *Telling Lies: Clues to Deceit in the Marketplace, Politics and Marriage*. New York: W.W. Norton & Co.

Emery, N.J. 2000. 'The eyes have it: the neuroethology, function and evolution of social gaze.' *Neuroscience and Biobehavioral Reviews* **24**, pp. 581–604.

Hoffman, D.D. 1998. *Visual Intelligence*. New York, London: W.W. Norton & Co. [Visual discrimination of objects reduced to a set of (numbered!) rules. An off-putting approach, but it pays off.]

Kreisler, H. 2004. 'Face to face: the science of reading faces.' Conversation with Paul Ekman on January 14 at the Institute of International Studies, UC Berkeley. Transcript available at www.globetrotter.berkeley.edu/people4/Ekman/ekman-cono.html

Miller, G.F. 1997. 'Protean primates: the evolution of adaptive unpredictability in competition and courtship.' In Whiten, A. & Byrne, R.W. eds *Machiavellian Intelligence II: Extensions and Evaluations*. Cambridge: Cambridge University Press. pp. 312–340.

O'Connell, S. 1997. *Mindreading: How We Learn to Love and Lie*. Oxford: Heinemann.

Tomasello, M. & Call, J. 1997. *Primate Cognition*. Oxford: OUP.

Six – Theories of vision

Al-Haythem, Ibn (Alhazen) (trans. IA Sabra) 1989. *The Optics of Ibn Al-Haytham*. London: The Warburgh Institute (University of London).

Al-Kindi, Abu. In Hozien, M. (ed.), 1993. *Islamic philosophy on-line*. Available from: www.muslimphilosophy.com/kindi/default.htm

El-Ehwany, A.F. 'Al-Kindi.' In Sharif, MM. (ed), [1963] 1983. *A History of Muslim Philosophy* 1. Karachi: Pakistan Philosophical Congress. p. 424.

Hoorn, W. v. 1972. *As Images Unwind*. Amsterdam: University Press Amsterdam. [Heavy-going but rewarding analysis of visual theory from the Greeks to Newton and Goethe.]

Ladurie, E. L-R. trans. Goldhammer, A. 1997. *The Beggar and the Professor: a Sixteenth-century Family Saga*. Chicago: University of Chicago Press. [A cheerful and fascinating history of the Platter family.]

Lindberg, D.C. 1976. *Theories of Vision: from Al-Kindi to Kepler*. Chicago: University of Chicago Press.

Sabra, A.I. 1981. *Theories of Light: from Descartes to Newton*. Cambridge: Cambridge University Press.

Wertheim, M. 1999. *The Pearly Gates of Cyberspace: a History of Space from Dante to the Internet*. New York, London: W.W. Norton & Co.

Seven – Nervous matter, visually endowed

Bentivoglio, M. 1998. 'Life and discoveries of Santiago Ramón y Cajal.' Stockholm: The Nobel Foundation. Available from www.nobelprize.org medicine/articles/cajal/index.html

Bruce, V., Green, P.R., Georgeson, M.A. 1996. *Visual Perception: Physiology, Psychology and Ecology* (3rd edition). Hove, E. Sussex: Psychology Press. [Vision science's very readable postgraduate-level bible.]

Cajal, S.R y. 1906 'The structure and connexions of neurons.' (Nobel lecture). In (Elsevier) 1967. *Nobel Lectures, Physiology or Medicine 1901–1921*, Amsterdam: Elsevier Publishing Company.

Hartline, H.K. 1967. 'Visual receptors and retinal interaction' (Nobel lecture). In (Elsevier) 1972. *Nobel Lectures, Physiology or Medicine 1963–1970*, Amsterdam: Elsevier Publishing Company. Available from: www.nobelprize.org/medicine/laureates/1967/hartline-lecture.html

Hartline, H.K. Wagner, H.G. and Ratliff, F. 1956. 'Inhibition in the eye of *Limulus*.' *The Journal of General Physiology* **39**, pp. 651–673.

Holmgren, E. In Grant, G. 1999. 'How Golgi shared the 1906 Nobel Prize in Physiology or Medicine with Cajal.' Available from www.nobelprize.org/medicine/articles/grant/index.html

Sharpe, L.T., Whittle, P., Nordby K. 1993. 'Spatial integration and sensitivity changes in the human rod visual system.' *Journal of Physiology* **461**, pp. 235–46.

The John Moran Eye Center at the University of Utah maintains 'Webvision: the organization of the retina and visual system', a thorough and colourful guide to the subject by Helga Kolb, Edurado Fernandez and Raph Nelson. Visit www.webvision.med.utah.edu/

Eight – Seeing Colours

Ball, P. 2002. *Bright Earth*. London: Penguin, [An erudite history of colour.]

Campbell, F.W. 1994. 'Dr. Edwin H. Land (1909–1991)' *Biographical Memoirs of Fellows of the Royal Society* **40**, pp. 195–219.

McElheny, V. 1998. *Insisting on the Impossible: the Life of Edwin Land*. Reading, Mass: Perseus Books.

Mollon, J.D. 2003. 'The origins of modern color science.' In Shevell, S. (ed) *Color Science*. Washington: Optical Society of America.

Newton, I. [1730/1952]. *Opticks, or a Treatise of the Reflections, Refractions, Inflections and Colours of Light*. New York: Dover.

Bruce MacEvoy's guide to watercolour painting includes a very good account of colour at www.handprint.com/hp/wcl/wcolor.html

The Virtual Colour Museum, a cultural history of colour written by Narciso Silvestrini and Ernst Peter Fischer, is available at www.colorsystem.com

Nine – Unseen colours

Crescitelli, F. 1963. 'Obituary: Gordon Lynn Walls (1905–1962).' *Vision Research* **3**, pp. 1–7.

Hunt, D.M., Dulai, K.S., Bowmaker, J.K., Mollon, J.D. 1995. 'The chemistry of John Dalton's color blindness.' *Science* **267**, pp. 984–8.

Hunt, D.M., Dulai, K.S., Cowing, J.A., Julliot, C., Mollon, J.D., Bowmaker, J.K., Li, W.H., Hewett-Emmett, D. 1998. 'Molecular evolution of trichromacy in primates.' *Vision Research* **38**, pp. 3299–3306.

Davenport, D.A. 1984. 'John Dalton's first paper and last experiments.' *Chem-Matters*, April 1984, p. 14.

Ramachandran, V.S. & Blakeslee, S. 1998. *Phantoms in the Brain*. London: Fourth Estate. [The humble blind spot harnessed to a dazzling account of consciousness and perception.]

Sacks, O. 1995. *An Anthropologist from Mars*. London: Pan Macmillan. [includes 'The Case of the Colourblind Painter.']

Sacks, O. 1996. *The Island of the Colourblind*. London: Pan Macmillan.

Zeki, S. 1980. 'The representation of colours in the cerebral cortex.' *Nature* **284**, pp. 412–418.

Ten – Making eyes to see

Brindley, G.S., & Rushton, D.N. 1974. 'Implanted stimulators of the visual cortex as visual prosthetic devices.' *Transactions of the American Academy of Ophthalmology & Otolaryngology* **78**, pp. 741–745.

Brooks, R.A. 2002. *Robot: the Future of Flesh and Machines*. London: Allen Lane. [A personal, sometimes sardonic take on the world of artificial intelligence. Among other accomplishments, Brooks has edited the *International Journal of Computer Vision*.]

Chua, C.N. 2000. 'Ophthalmology in the British Isles.' Available from: www.mrcophth.com/Historyofophthalmology/ophthalmologyinuk.htm

Dobelle, W.H. 2000. 'Artificial vision for the blind by connecting a television camera to the visual cortex.' *Transactions – American Society for Artificial Internal Organs* **46/1**, pp. 3–9.

Dobelle, W.H., Quest, D.O., Antunes, J.L., Roberts, T.S., Girvin, J.P. 1979. 'Artificial vision for the blind by electrical stimulation of the visual cortex.' *Neurosurgery* **5**, pp. 521–7.

Gregory, R.L. & Wallace, J.G. 1963. *Recovery from early blindness – a case study.* (Experimental Psychology Society Monograph No. 2) Available from: www.richardgregory.org/papers/index.htm

Lindblom, J. & Ziemke, T. 2003 'Social situatedness of natural and artificial intelligence: Vygotsky and beyond.' *Adaptive Behavior* **11/2**.

Rizzo, J.F., *et al.* 'Retinal prosthesis: an encouraging first decade with major challenges ahead.' *Ophthalmology* **108/1**, pp. 13–14.

Epilogue – The invisible gorilla

Noë, A., Pessoa, L. & Thompson, E. 2000. 'Beyond the grand illusion: what change blindness really teaches us about vision.' *Visual Cognition* **7**, p. 93.

Simons, D.J. & Levin, D.T. 1998. 'Failure to detect changes to people during real-world interaction.' *Psychonomic Bulletin and Review* **4**, p. 644.

Spinney, L. 2000. 'How much of the world do we really see?' *New Scientist* **2265**.

The Visual Cognition Lab at Illinois University maintains footage of attention blindness experiments at www.viscog.beckman.uiuc.edu/djs_lab/demos.html

NOTES

Prologue – Youth and age

1 Gordon, M. 1997. 'Unravelling the real story behind the Cyclops.' *New Scientist*, **153/2068**, p. 16.

2 Gould, G.M. & Pyle, W.L. [1896] 1997. *Anomalies and Curiosities of Medicine.* Seattle, WA: The World Wide School. Available at: www.gutenberg.org/ etext/747

3 Maurice, D.M. 1998. 'An ophthalmological explanation of REM sleep.' *Experimental Eye Research* **66**, pp. 139–145.

4 Siegel, J.M. 2005. 'Functional implications of sleep development.' *PLoS Biology* **3/5**, p. 178.

5 Tarusov, B.N., Polivoda, A.I., Zhuravlev, A.I. 1961. 'Study of the faint spontaneous luminescence of animal cells.' *Biophysics* **6**, pp. 83–85.

6 Cepko, C. Interview with Norman Swan on *The Health Report: the Retina and the Brain*, ABC Radio National, Australian Broadcasting Commission, Monday 13 December 1999. Transcript available at www.abc.net.au/rn/ talks/8.30/helthrpt/stories/s73272.htm
Livesey F.J., Cepko C.L. 2001. 'Vertebrate neural cell-fate determination: lessons from the retina.' *Nature Reviews: Neuroscience* **2/2**, pp. 109–18.

7 The distinction between rods and cones is apparent in the jawless fish of the Devonian Period, 417–354Ma. See Bowmaker, J.K. 1991. 'Evolution of

photoreceptors and visual pigments.' In J.R. Cronly-Dillon and R.L. Gregory, eds. *Evolution of the Eye and Visual Pigments*. Boca Raton, Fla: CRC Press, pp. 63–81

8 Schefrin, B.E., Bieber, M.L., McLean, R., Werner, J.S. 1998. 'The area of complete scotopic spatial summation enlarges with age.' *Journal of the Optical Society of America A (Optics, Image Science and Vision)* 15/2, pp. 340–8. But see Schefrin, B.E., Hauser, M., Werner, J.S. 2004. 'Evidence against age-related enlargements of ganglion cell receptive field centers under scotopic conditions.' *Vision Research*, 44/4, pp. 423–8.

One – The commonwealth of the senses

1 Keller, H. [1905] 2002. *The Story of My Life*. New York: Penguin Books

2 Yarbus, AL. 1967. 'Eye movements during perception of complex objects', in L.A. Riggs, ed, *Eye Movements and Vision*. New York: Plenum Press, pp. 171–196.

3 Dennett, D. 1991. *Consciousness Explained*. New York: LittleBrown & Company, pp. 339–342.

4 Catania, K.C. 1999. 'A nose that looks like a hand and acts like an eye: the unusual mechanosensory system of the star-nosed mole.' *Journal of Comparative Physiology, A: Sensory, Neural, and Behavioral Physiology* 185, pp. 367–372.

5 Wolken, J.J. & Florida, R.G. 1969. 'The eye structure and optical system of the crustacean copepod *Copilia*'. *Journal of Cell Biology* 40/1, pp. 279–85.

6 O'Shea, S. 2003. Letter to *the Octopus News Magazine online* message board, 24 May. Available at www.tonmo.com/forums

7 Milius, S. 2003. 'Moonlighting: beetles navigate by lunar polarity.' *Science News* 164/1, p. 4.

8 Shashar, N. & Cronin, T.W. 1996. 'Polarization contrast in octopus.' *The Journal of Experimental Biology* 199, pp. 999–1004.
Shashar, N., Rutledge, P.S. & Cronin, T.W. 1996. 'Polarization vision in cuttle-fish – a concealed communication channel?' *The Journal of Experimental Biology* 199, pp. 2077–2084.

9 Adapted from Vilhjalmsson, T. 1997. 'Time and travel in Old Norse society.' *Disputatio* 2/89, pp. 114.

10 Dowling, J.E. 2000. 'George Wald, November 18, 1906 – April 12, 1997.' *Biographical Memoirs* 78. National Academy of Sciences Press. Available at www.nap.edu/html/biomems/gwald.html

11 Chiao, C-C., Cronin, T.W. & Osorio, D. 2000. 'Color signals in natural scenes: characteristics of reflectance spectra and effects of natural illuminants.' *Journal of the Optical Society of America A* **17**, pp. 218–224.

12 Morgan, M.J., Adam, A., Mollon, J.D. 1992. 'Dichromats detect colour-camouflaged objects that are not detected by trichromats.' *Proc. R. Soc. Lond. B Biol. Sci* **248/1323** pp. 291–5.

13 Marshall, N.J. 2000. 'Communication and camouflage with the same "bright" colours in reef fishes.' *Philosophical Transactions of the Royal Society of London B* **355**, pp. 1243–1248. Available at www.dx.doi.org/10.1098/rstb.2000.0676

Milius, S. 2004. 'Hide and see: conflicting views of reef-fish colors.' *Science News* **166/19**, p. 296. Available at http://www.sciencenews.org/articles/20041106/bob8.asp

14 'Bullet Shrimp.' *Mirror*, Friday April 10, 1998, p. 11.

15 Chiao, C-C., Cronin, T.W., Marshall, N.J. 2000. 'Eye design and color signaling in a stomatopod crustacean *Gonodactylus smithii*.' *Brain, Behavior and Evolution* **56**, pp. 107–122.

16 Leonardo da Vinci. In Wheatstone, C. 1838.

17 Wheatstone, C. 1838. 'Contributions to the physiology of vision 1: on some remarkable, and hitherto unobserved, phenomena of binocular vision.' *Philosophical Transactions of the Royal Society of London* **128**, pp. 371–394.

18 Julesz, B. 1995. *Dialogues on Perception*. Cambridge, Massachusetts: MIT Press.

19 Of course, producing a pattern of truly random dots on a piece of paper – without the aid of a computer – would be the truly heroic part of such an enterprise. Bela Julesz used an IBM-704 computer to generate his patterns on a tape which could then be converted into a television signal. See Mollon, J.D. 1997. '"... On the basis of velocity clues alone": some perceptual themes 1946–1996,' *The Quarterly Journal of Experimental Psychology*, **50A/4**, pp. 859–878.

20 Hartline, H.K. 1972. 'Visual receptors and retinal interaction.' In *Nobel Lectures, Physiology or Medicine 1963–1970*, Elsevier Publishing Company, Amsterdam p281. Available at www.nobelprize.org/medicine/laureates/1967/hartline-lec

21 The quotations here are taken from Wade, N.J. 2005. 'The original spin doctors – the meeting of perception and insanity.' *Perception* **34**, pp. 253–260.

22 Hubel, D.H. 1988. *Eye, Brain and Vision*. New York: Scientific American Library, p. 81.

23 Oyster, C.W. 1999, p. 178.

24 Gilchrist, I.D., Brown, V., Findlay, J.M. 1997. 'Saccades without eye movements.' *Nature* **390**, pp. 130–1.

25 Weiskrantz, L. 1986. *Blindsight: a Case Study and Implications*. Oxford:
Clarendon Press.

26 MacKenzie, D. 1999. 'An eye for danger.' *New Scientist* **2188**, p. 21.

Two – The chemistry of vision

1 Oomen, H.A. 1958. 'Clinical experience on hypovitaminosis A.' In Kinney,
T.D. & Follis, R.H. Jr 1958. *Nutritional disease*. Federation Proceedings **17**,
Suppl 2, No 3, Pt 2, pp. 111–128.

2 Campion-Vincent, V. 1999. 'The tell-tale eye.' *Folklore* **110**, pp. 13–24.

3 de l'Isle-Adam, V. (trans. Symons, A) 1925. *Claire Lenoir*. New York: Albert
& Charles Boni.

4 Goulet, A. 2005. 'South Sea daggers and the dead man's eye: foreign
invasion in *fin-de-siecle* optogram fiction.' *Cahiers Victoriens et Édouardiens* **61**.

5 Boll, F. 1877. 'Zur anatomie und physiologie der retina.' *Arch. Anat. Physiol.*
(Physiol. Abt) **4–35**.

6 Wald, G. 1953. 'Eye and camera.' In *Scientific American Reader*. New York:
Simon & Schuster. pp. 555–68.

7 Vernois, M. 1870. 'Etude photographique sur la retine des sujets assassines.'
Revue photographique des hopitaux de Paris **2**, pp. 73–82. In Campion-
Vincent, 1999.

8 Dew, W. 1938. *I caught Crippen*, London: Blackie & Son.

9 Magendie, F. In Grmek, M.D. (1974) 'François Magendie.' *Dictionary of scientific biography* **9**, pp. 6–11.

10 Adapted from Boyle, R. In Hunter, M. 2004. *The Robert Boyle Project*. Available at: www.bbk.ac.uk/boyle.

11 Wolf, G. 1978. 'A historical note on the mode of administration of vitamin
A for the cure of night blindness.' *American Journal of Clinical Nutrition*
31, pp. 290–2.

12 Celsus, A.C. In McLaren, D.S. [1] *The control of xerophthalmia a century
of Contributions and Lessons*. Basel: Sight and Life. Available at: www.sightan-
dlife.org/booksAll/BooksHTML/allBooks.html

13 Elliot, R.H. 1920. *Tropical Ophthalmology*. London: Oxford Medical Publi-
cations. p. 78.

14 Wald, G. 1968. 'The molecular basis of visual excitation.' In Granit R. (ed). *Les Prix Nobel en 1967*. Stockholm: Nobel Foundation.

Three – How are eyes possible?

1 Axelrod, J. 1988. 'An unexpected life in research.' *Annual review of pharmacology and Toxicology* **28**, pp. 1–23.

2 Masland, R.H. 2001. 'The fundamental plan of the retina.' *Nature Neuroscience* **4/9**, pp. 877–886.

3 Okano, T., Kojima, D., Fukada, Y. & Shichida, Y. 1992. 'Primary structures of chicken cone visual pigments: vertebrate rhodopsins have evolved out of cone visual pigments.' *Proceedings of the National Academy of Sciences of the United States of America* **89** pp. 5932–5936.
  Okano, T. 2002. 'Pinopsin is a chicken pineal photoreceptive molecule' (letter). *Nature* **372**, pp. 94–97.
  Su, C-Y. 2006. 'Parietal-eye phototransduction components and their potential evolutionary implications.' *Science* **311/5767**, pp. 1617–1621.

4 Paley, W. [1854] 2006. *Natural Theology*. New York, Oxford: Oxford University Press.

5 Gehring, W.J. 1998. *The Homeobox Story*. New Haven, CT: Yale University Press. p. 207.

6 Oyster, C.W. 1999, p. 71.

7 Angier, N. 1995. 'With new fly, science outdoes Hollywood.' *New York Times* March 24: A1 (col. 2), A15 (col. 1).

8 Halder, G., Callaerts, P., Gehring W.J. 1995. 'Induction of ectopic eyes by targeted expression of the eyeless gene in Drosophila.' *Science* **267**, pp. 1788–92.

9 Dickinson, W.J. & Seger, J. (response, Gehring, W.J.) 1996. 'Letters: eye evolution.' *Science* **272**, pp. 467–471.

10 Travis, J. 1997. 'Eye-opening gene.' *Science News* **151/19**.

11 Goethe, J.W. In Seamon, D. & Zajonc, A. 1998. *Goethe's Way of Science*. New York: State University of New York Press, p. 199.

12 See Conway-Morris, S. & Gould, S.J. 'Showdown on the Burgess Shale.' *Natural History* **107/10**, pp. 48–55.

13 Parker, A. 2003. *In the Blink of an Eye*. London: Perseus Books.

14 Conway-Morris, S. 2003. 'On the first day, God said . . .' *American Scientist* **91/4**.

15 (quotation) Allen, G. In Mollon, J.D. 2003.

Four – The adaptable eye

1 Nilsson, D-E. 1989. 'Optics and evolution of the compound eye.' *Facets of vision* (Stavenga, D.G. & Hardie, R.C. eds). Berlin-Heidelberg: Springer-Verlag. p. 3075.

2 Mollon, J.D. 1997. '"... On the basis of velocity clues alone": some perceptual themes 1946–1996'. *The Quarterly Journal of Experimental Psychology*, **50A/4**, pp. 859–878.

3 See, for example, Thakoor, S., Morookian, J.M., Chahl, J., Hine, B., Zornetzer, S. 2004. 'BEES: Exploring Mars with Bioinspired Technologies.' *Computer*, **37/9**, pp. 38–47.

4 Rainer Schönhammer, R. 2000. 'Flying (human) bodies in the fine arts – dreams and daydreams of flying.' Paper presented at the 17th Annual International Conference of the Association for the Studies of Dreams, July 4–8, 2000, Washington, DC. Available from: www.psydok.sulb.uni-saarland.de/volltexte/2005/550/

5 Kelber, A., Balkenius, A., Warrant, E.J. 2002. 'Scotopic colour vision in nocturnal hawkmoths' (letter). *Nature* **419**, pp. 922–925.

6 Gould, G.M. & Pyle, W.L. [1896] 1997.

7 Boyce, N. 2000. 'Eye wish.' *New Scientist* **167/2253**, p. 33.

Five – Seeing and thinking

1 Lorenz, K. 1992. 'Analogy as a source of knowledge.' *Nobel Lectures, Physiology or Medicine 1971–1980*, (Lindsten, J. ed) Singapore: World Scientific Publishing Co.

2 Helmholtz, H. v. 1860 *Treatise on Physiological Optics, vol. 3*. In Watson, R.I. (ed.). 1979. *Basic Writings in the History of Psychology*. New York: Oxford University Press.

3 Millard, R. 2002. *The Tastemakers*. London: Scribner.

4 Chance, M.R.A. 1957. 'The role of convulsions in behavior.' *Behavioral Science* **2**, 30–45.

5 Tiger, L. 'Real-life survivors rely on teamwork.' *Wall Street Journal* August 25, 2000, p. A14.

6 Miller, H. 1991 'Eye contact.' Letter to *New Scientist* **1781**, p. 57.

7 O'Connell, S. 1997. *Mindreading: how we learn to love and lie*. Oxford: Heinemann, p. 60.

8 Thompson, P. 1980. 'Margaret Thatcher: a new illusion.' *Perception* **9**, pp. 483–484.

9 Kreisler, H. 2004. 'Face to face: the science of reading faces.' Conversation

with Paul Ekman on January 14 at the Institute of International Studies, UC Berkeley. Transcript available at www.globetrotter.berkeley.edu/people4/ Ekman/ekman-cono.html

10 Brockman, J. 1999. 'Animal minds: a talk with Marc D. Hauser.' *Edge* **54**. Available at: www.edge.org/documents/archive/edge54.html

Six – Theories of vision

1 Zajonc, A. 1993. *Catching the Light*. Oxford: Oxford University Press.

2 Al Kindi. In Hozien, M. (ed.), 1993 *Islamic philosophy on-line*. Available at: www.muslimphilosophy.com/kindi/default.htm

3 Deregowski, J.B. 1972. 'Pictorial perception and culture.' *Scientific American* **227**, pp. 82–88.

4 See, for example, Borgmann, A. 1999. *Holding on to Reality: the Nature of Information at the Turn of the Millennium*. Chicago: University of Chicago Press.

5 Hoffman, 1998, p. 199.

6 Kepler, J. In Arthur Koestler, 1989. *The Sleepwalkers*. London: Penguin Books, p. 280.

7 Platter, F. 1583. *De corporis humani structura et usu*. In Lindberg, 1976, p. 176.

8 Kepler, J. 1604. *Ad Vitellionem paralipomena*. Translated in Donahue, W.H. 2000. *Optics: Paralipomena to Witelo and the optical part of astronomy*. pp. 151–2.

Seven – Nervous matter, visually endowed

1 Purkinje, J. In Wyman, M. 2001. 'The history of veterinary ophthalmology with particular emphasis for Ohio.' 25th annual Waltham/OSU Symposium on Small Animal Ophthalmology. Available at www.vin.com/VINDBPub/ SearchPB/Proceedings/PR05000/PR00525.htm

2 Campbell, L. & Garnett, W. 1882. *The Life of James Clerk Maxwell*. London: Macmillan & Co. Digital preservation © 1997, 1999 Sonnet Software, Inc. Available at www.sonnetusa.com/bio/maxwell.asp

3 Keeler, C.R. 2004. 'Babbage the unfortunate.' *British Journal of Ophthalmology* **88**, pp. 730–732.

4 Helmholtz, H v. (trans. E. Atkinson) 1999. *Popular Lectures on Scientific Subjects, Second Series*. Bristol: Thoemmes Press. p. 278.

5 Brewster, D. 1832. In Ramachandran, V.S. & Gregory, R.L. 1991.

6 Schickore, J. 1998. 'The Microscopic Anatomy of the Retina, 1835–1855.' Paper presented at the History of Science Society Annual Meeting, October 23. Missouri: Kansas City.

7 James Clerk Maxwell's letter to his wife Katherine, January 3, 1870. In Campbell, L. & Garnett, W. 1882. p. 241.

8. Letter from William Swan, April 2 1882. In Campbell, L. & Garnett, W. 1882. pp. 236–7.

9 Retzius, G. trans. Grant, G. 1948. *Biografiska Anteckningar och Minnen* Vol. **2**. Uppsala: Almqvist & Wiksell, p. 246.

10 Holmgren, E. In Grant, G. 1999. 'How Golgi Shared the 1906 Nobel Prize in Physiology or Medicine with Cajal.' Available from www.nobelprize.org/medicine/articles/grant/index.html

11 Bernhard, C.G. 1967. Presentation speech, the Nobel Prize in Physiology or Medicine. In Granit, R. (ed) 1968. *Les Prix Nobel en 1967*, Stockholm: Nobel Foundation.

12 Dakin, S.C. & Bex, PJ. 2003 'Natural image statistics mediate brightness "filling in".' *Proc R Soc Lond B Biol Sci.* **270**, pp. 2341–8. Dakin and colleagues at University College, London are now conducting brain-scans to try and isolate the 'filling in' mechanism triggered by his illusion.

13 Sharpe, L.T., Whittle, P. Nordby, K. 1993. 'Spatial integration and sensitivity changes in the human rod visual system.' *Journal of Physiology* **461**, pp. 235–46.

Eight – Seeing colours

1 Mollon, J.D. 2003 'The origins of modern color science.' In Shevell, S. (ed) *Color Science.* Washington: Optical Society of America.

2 Gladstone, W. 1877 'The colour sense.' *Nature* **19thC.2** pp. 366–88.

3 Geiger, L. In Bellmer, E.H. 1999. 'The statesman and the ophthalmologist: Gladstone and Magnus on the evolution of human colour vision, one small episode of the nineteenth-century Darwinian debate.' *Annals of Science* **56/1**, pp. 25–45.

4 Gage, J. 1993. In Ball, P. 2002. *Bright Earth*. London: Penguin, p. 49.

5 Newton, I. 1671/72 'A Theory Concerning Light and Colors.' *Philosophical Transactions of the Royal Society of London* **80**. Available at: www.newton-project.ic.ac.uk/texts/cul3970_n.html

6 Newton, I. [1730] 1952. *Opticks, or a Treatise of the Reflections, Refractions, Inflections & Colours of Light.* 4th edition. New York: Dover.

7   Knight, D. 2002. 'Scientific lectures: a history of performance.' *Interdisciplinary Science Reviews.* **27/3**, pp. 217–224.

8   Young, T. 1804. 'Experimental demonstration of the general law of the interference of light.' *Philosophical Transactions of the Royal Society of London* **94**.

9   Campbell, F.W. 1994. 'Dr Edwin H. Land (1909–1991).' *Biographical Memoirs of Fellows of the Royal Society* **40**, pp. 195–219.

10 Mortimer, C. 1731. 'An account of Mr Christopher Le Blon's principles of printing, in imitation of painting and of weaving tapestry, in the same manner as Brocades.' *Philosophical Transactions of the Royal Society of London*, **37**, pp. 101–7. In Mollon, J.D. 2003.

11 Julesz, B. 1995. *Dialogues on Perception.* Cambridge, Mass: MIT Press. p. 27.

Nine – Unseen colours

1   Dalton, J. In Davenport, DA. 1984. 'John Dalton's first paper and last experiments.' *ChemMatters*, April 1984, p. 14.

2   Frey, F.G. 1975. 'A railway accident a hundred years ago as reason for systematic testing of colour vision.' *Klinische Monatsblätter für Augenheilkunde* **167/1** pp. 125–7.

3   Zorpette, G. 2000. 'Looking for Madam Tetrachromat.' *Red Herring*, November 1.

4   Hubel, 1988. p. 165.

5   Wilson, E.B. In Al-Awqati, Q. 'Edmund Beecher Wilson, America's first cell biologist.' *Columbia magazine*, Fall 2002. Available at www.columbia.edu/cu/alumni/Magazine/Fall2002/Wilson.html

6   Jaeger, W. 1992. 'Horner's law. The first step in the history of the understanding of X-linked disorders.' *Ophthalmic paediatrics and genetics* **13/2**, pp. 49–56.

7   Hunt, D.M., Dulai, KS., Bowmaker, J.K., Mollon, J.D. 1995. 'The chemistry of John Dalton's color blindness.' *Science* **267**, pp. 984–8.

8   Jameson, K.A., Highnote, S.M., Wasserman, L.M. 2001. 'Richer color experience in observers with multiple photopigment opsin genes.' *Psychonomic Bulletin & Review* **8/2**, pp. 244–261.

9   Flom, M.C., Stern, C., White, H.E. 1963. 'University of California: *in memoriam* Gordon Lynn Walls, Optometry; Physiology: Berkeley.' Available at: www.crsltd.com/research-topics/walls/obituary.html

10 Roth, L. & Kelber, A. 2004. 'Colour vision in nocturnal and diurnal geckos.' The Lund Vision Group, Lund University. Available at: www.biol.lu.se/funkmorf/vision/almut/gecko/html

11 Arrese, C., Hart, N., Thomas, N., Beazley, L., Shand, J. 'Trichromacy in Australian Marsupials.' *Current Biology* **12/8**, pp. 657–660.

12 Huth, G.C. 'A new physics-based model for light interaction with the retina of the eye: rethinking the vision process.' Available at www.ghuth.com/vision/?p=55

13 Cottaris, N.P. & de Valois, R.L. 1998. 'Temporal dynamics of chromatic tuning in macaque primary visual cortex.' *Nature* **395**, pp. 896–900.

14 Smithson, H.E. & Mollon, J.D. 2004. 'Is the S-opponent chromatic subsystem sluggish?' *Vision Research* **44**, pp. 2919–2929.

15 Nathans, J., Thomas, D., & Hogness, D.S. 1986. 'Molecular genetics of human color vision: the genes encoding blue, green, and red pigments.' *Science* **232**, pp. 193–202.

16 Mollon, J.D. 1999. 'Color vision: opsins and options.' *Proceedings of the National Academy of Sciences of the United States of America* **96**, pp. 4743–4745.

17 Sacks, O. 1995. *An Anthropologist from Mars: Seven Paradoxical Tales.* London: Picador.

18 Julesz, 1995, p. 47.

19 Mollon, J.D. 1997. '"... On the basis of velocity clues alone": some perceptual themes 1946–1996'. *The Quarterly Journal of Experimental Psychology*, **50A/4**, p. 867.

20 Mach, E. [1886] 1987. *Contributions to the Analysis of the Sensations.* Peru, IL: Open Court.

Ten – Making eyes to see

1 Bedard, E.R. 2002. 'Non-lethal capabilities: realizing the opportunities.' *Defense Horizons* **9**.

2 Burgess, L. 'Troops train to use non-lethal weapons to control crowds, reduce civilian deaths.' *Stripes Sunday magazine*, 22 December 2002.

3 Rappert, B. 2003. *Non-lethal Weapons as Legitimising Forces? Technology, Politics and the Management of Conflict.* London: Frank Cass Publishers.

4 Hambling, D. 2002. '"Safe" laser weapon comes under fire.' *New Scientist*, **2359**, p. 5.

5 Dunham, W. 2006. 'Pentagon uses laser device at Iraqi checkpoints.' *Reuters* 18 May, 17:13:42 GMT.

6 Adapted from Chau, C.N. 2000. 'Ophthalmology in the British Isles.' Available from:www.mrcophth.com/Historyofophthalmology/ophthalmologyinuk.htm

7 Locke, J. [1690] 2004. *Essay Concerning Human Understanding* 2/9.8. London: Penguin Books.

8 Gregory, R.L. & Wallace, J.G. 1963. *Recovery from early blindness – a case study*. (Experimental Psychology Society Monograph No. 2) Available from: www.richardgregory.org/papers/index.htm

9 See, for example, Snyder, F.W. & Pronko N.H. 1952. *Vision with spatial inversion*. Wichita, KS: McCormick-Armstrong – a detailed and entertaining account.

10 Julesz, B. 1995. *Dialogues on Perception*. Cambridge, Massachusetts: MIT Press, p. 3.

11 'Development of biomorphic flyers.' *NASA tech briefs* November 2004, p.54. Summary available from: www.nasatech.com/Briefs/Nov04/NPO30554 .html

12 Lindblom, J. & Ziemke, T. 2003. 'Social situatedness of natural and artificial intelligence: Vygotsky and beyond.' *Adaptive Behavior* 11/2.

13 Vygotsky, L. S. 1979. 'Consciousness as a problem in the psychology of behaviour.' *Soviet Psychology* 17/4, pp. 3–35.

14 See, for example, Trevarthen, C. 'Can a robot hear music? Can a robot dance? Can a robot tell what it knows or intends to do? Can it feel pride or shame in company? Questions to the nature of human vitality.' In Prince, C.G., Demiris, Y., Marom, Y., Kozima, H., Balkenius, C. (eds) 2002. *Proceedings of the Second International Workshop on Epigenetic Robotics: modeling cognitive development in robotic systems* (Lund University Cognitive Studies 94). Lund: Lund University, pp. 79–86.

15 Wagenaar, D.A. 2004. 'Cortical stimulation for the evocation of visual perception.' Term paper for CNS247: Cerebral cortex, R. Andersen, Caltech. Available from: www.its.caltech.edu//pinelab/wagenaar/papers/cns247.pdf

16 Brindley, G.S. & Lewin, W.S. 1968. 'The sensations produced by electrical stimulation of the visual cortex.' *The Journal of Physiology* 196, pp.479–493.

17 See also Dobell, W.H., Mladejovsky, M.G. 1974. 'The directions for future research on sensory prostheses.' *Transactions – American Society for Artificial Internal Organs* 20, pp. 425–429.

18 Heiduschka, P. & Thanos, S. 1998. 'Implantable bioelectric interfaces for lost nerve functions.' *Progress in neurobiology* 55, pp. 433–61.

19 Humayun, M.S., Sato, Y., Propst, R. & de Juan Jr. E. 1995. 'Can potentials from the visual cortex be elicited electrically despite severe retinal degeneration and a markedly reduced electroretinogram?' *German Journal of Ophthalmology* **4**, pp. 57–64.

20 Ananthaswamy, A. 2002. 'Eye strain.' *New Scientist* **2329**, p. 14.

21 Wheatstone, C. 1838.

22 'Computer Helps Blind Man "See".' Reuters 09:20 AM Jan. 17, 2000 PT. Available from: www.wired.com/news/technology/0,1282,33691,00.html

Epilogue – The invisible gorilla

1 Spinney, L. 2000. 'Blind to change' *New Scientist* **2265**, pp. 28–32.

PICTURE CREDITS

PLATES

The compound eye of the Antarctic krill *Euphausia superba* (Gerd Alberti
    and Uwe Kils)
The carapace-eye of a brittlestar (in Fitzgerald, R., 2001 © American
    Institute of Physics)

The eye of a decapod shrimp (Michael Land)
Close-up of the surface of a crayfish eye (Michael Land)

The mantis shrimp (Jeffrey Jeffords)
David Brewster's study of the visible spectrum (Brewster, D., 1831)

Some flowers boast ultraviolet patterns (Thomas Eisner, Cornell University)
The Banded Butterflyfish *Chaetodon striatus* (Tom Doeppner)

Slits, horseshoes and pinholes: pupil shapes are astonishingly various
    (Dan-Eric Nilsson)

The compound eye of *Phacops* (Riccardo Levi-Setti)
Erbenochile (Phil Crabb, courtesy of the Natural History Museum, London)

A star-nosed mole's nose (Kenneth Catania)
A naked molerat's teeth (Kenneth Catania)

The nerve cells of the retina, drawn by Santiago Ramón y Cajal (Instituto Cajal)

INDEX

A NOTE ON THE AUTHOR

Simon Ings is a novelist and science writer. *The Eye* was written in between the birth of his daughter and film-making expeditions to Ladakh, Arabia's Empty Quarter, and Arctic Norway. His science features and interviews have featured on national radio and in magazines as diverse as *New Scientist*, *Wired*, and *Dazed and Confused*. He lives in London.